LIFE IN THE TAR SEEPS

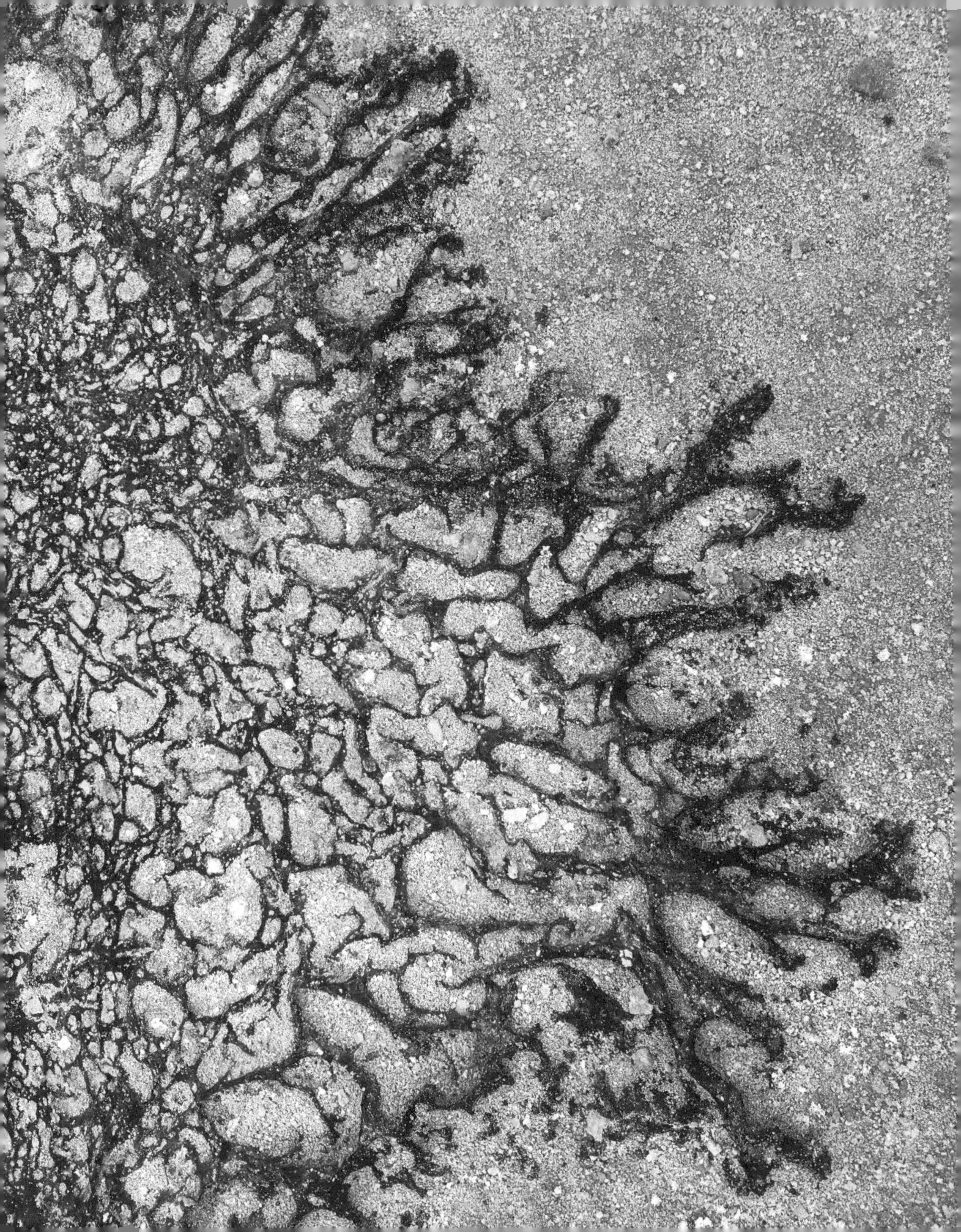

Life in the Tar Seeps

A Spiraling Ecology from a Dying Sea

Gretchen Ernster Henderson

TERRA FIRMA BOOKS /
TRINITY UNIVERSITY PRESS
SAN ANTONIO

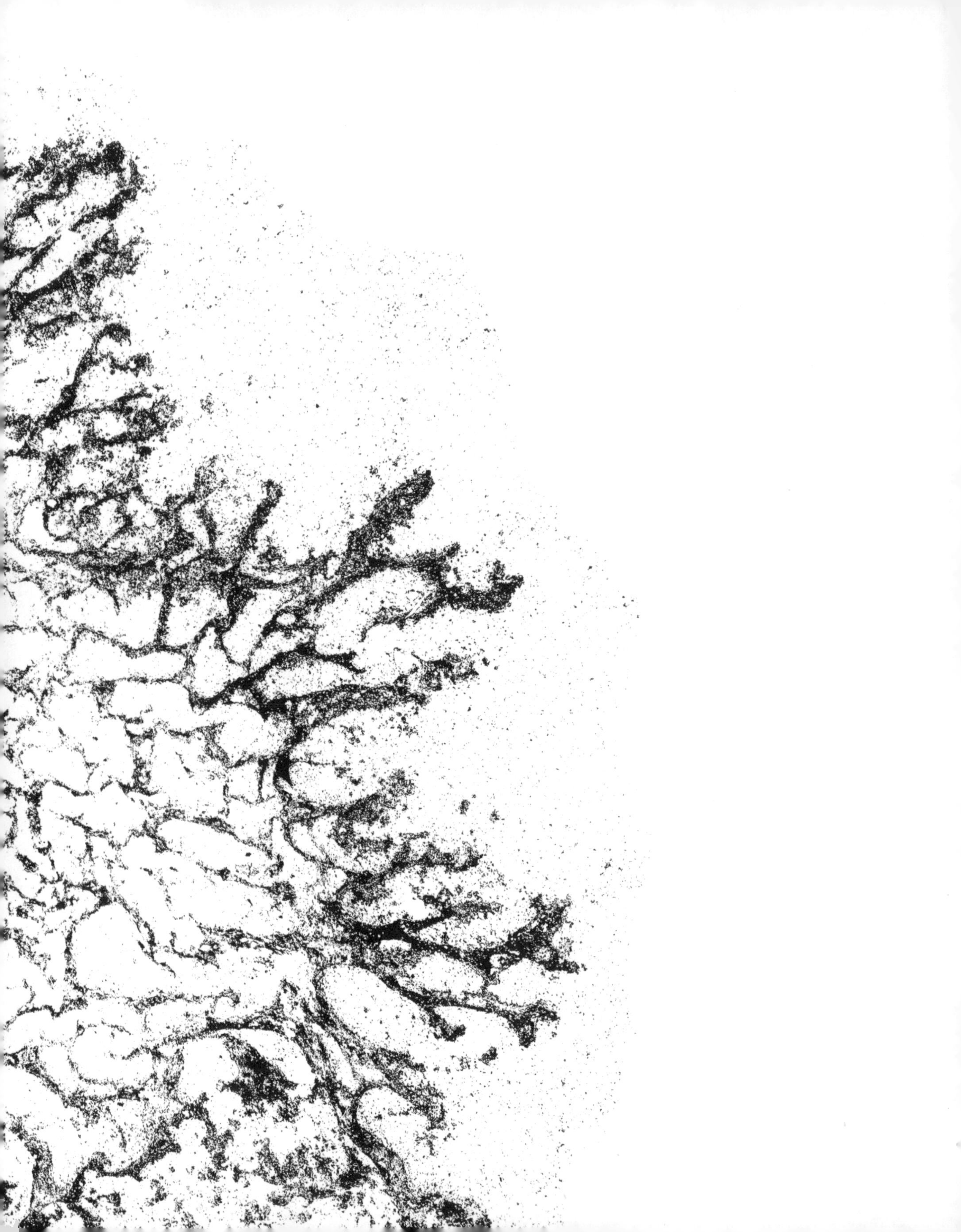

To Riley & Landon
& the Future
(that is now)

The tar cools . . .
The strata of the Earth is a jumbled museum.
Embedded in the sediment is a text.
—**Robert Smithson**, "A Sedimentation of the Mind"

The birds know better . . .
descending in majestic spirals . . .
—**Aldo Leopold**, *A Sand County Almanac*

I've heard it in the chillest land—
And on the strangest Sea—
—**Emily Dickinson**, "'Hope' is
the thing with feathers—"

The whole process . . . seems
caught up in an endless spiral.
—**Rachel Carson**, *Silent Spring*

CONTENTS

WAYFINDING

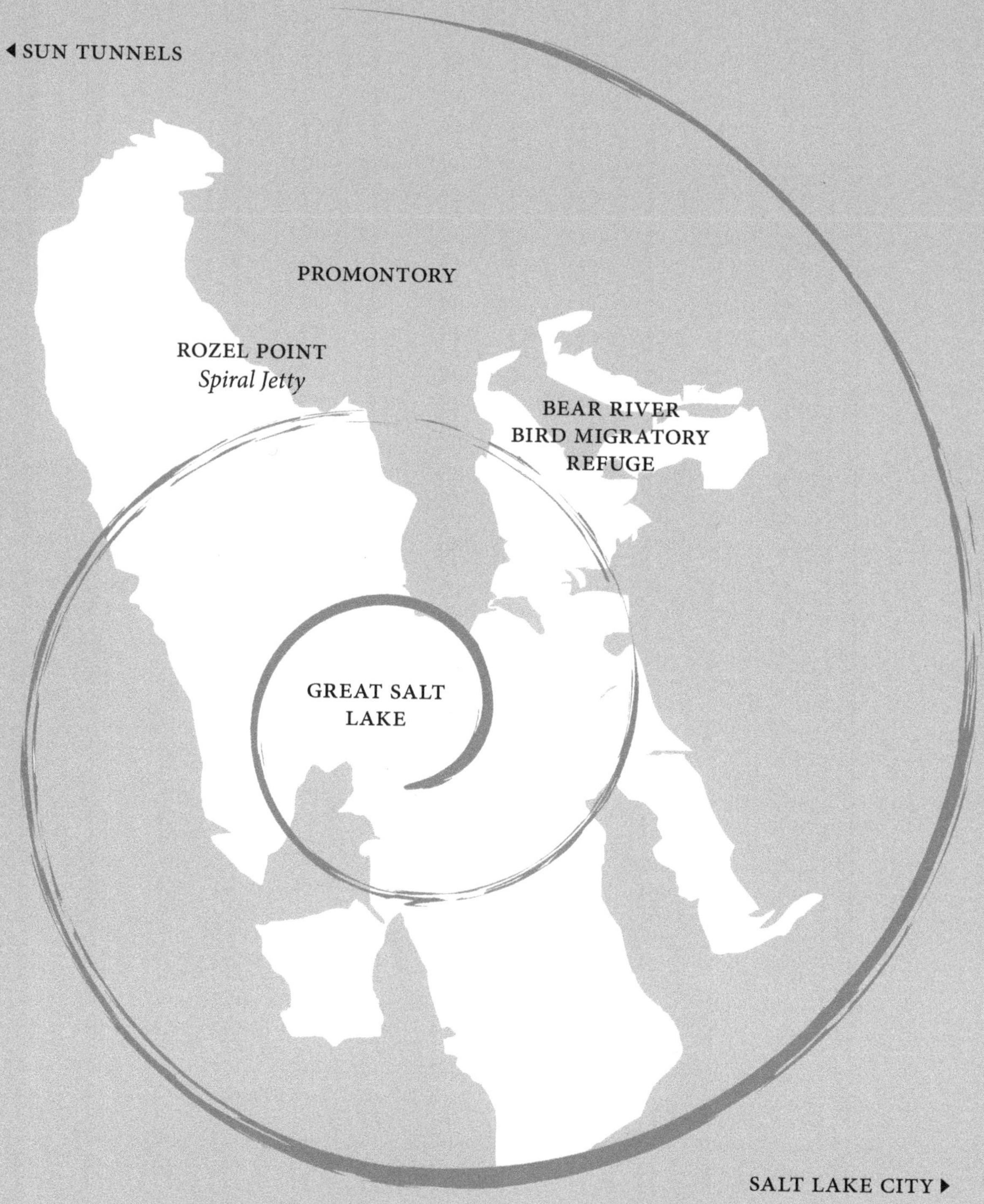

◂ SUN TUNNELS

PROMONTORY

ROZEL POINT
Spiral Jetty

BEAR RIVER
BIRD MIGRATORY
REFUGE

GREAT SALT
LAKE

SALT LAKE CITY ▸

PREFACE: Composition

I went to Great Salt Lake after recovering from being hit by a car in a crosswalk—on manmade asphalt—but it took me longer to correlate the lake's tar seeps—of natural asphalt—by comparison. Nicknamed "death traps," tar seeps are pools of raw oil that creep up from tectonic fractures and spread across the earth like sticky flypaper. An unsuspecting animal that crosses a melting seep can get fatally stuck.

In Great Salt Lake's remote north arm, at Rozel Point, a number of tar seeps have surfaced as the lake retreats from drought. Over a few years I witnessed a team of environmental scientists, artistic curators, land managers, and students working collaboratively to steward this challenging place. As I visited the lake again and again, life and death, degeneration and regeneration, injury and healing slowly started to congeal. My accident colored the backdrop against which I came to see the lake—not as dead but as wildly alive—a watershed for shifting perceptions of any overlooked place.

Great Salt Lake is a vibrant, living body of water that supports many lives. Often dismissed as a dead sea, the lake's story shifts with water cycles, bird migrations, microbial studies, environmental arts, and cultural histories shaped by Indigenous knowledges, overwritten by colonial settlements whose legacies live on in environmental threats. As a fifth-generation Californian who has lived many years on the East Coast, I went to Utah for two years as a visiting professor of environmental humanities who grew to wonder why the region's namesake was often dismissed as "stinky" and "ugly": virtually hiding in plain sight. As I navigated varied ecotones *(a transition zone between communities containing characteristic species; a place of danger or opportunity; a testing ground),* the tar seeps stuck together often-separated matters elsewhere and here.

In this desolate and spare landscape in the high desert—a place of surreal, ugly beauty—many convergences occur around the tar seeps, strikingly marked by Robert Smithson's iconic 1970 Land Art called *Spiral Jetty*. The artist's massive coil of salt-encrusted, black basalt unfurls into the lake—three times counterclockwise, around fifteen feet wide and fifteen hundred feet long—just down shore from the tar seeps. Smithson selected Rozel Point because of the natural tar seeps near abandoned attempts at oil drilling. Some first-time visitors even mistake the straight jetty seeping with raw oil for his spiraling artwork. The close proximity invites comparison.

Through repeated visits, I grew to perceive *Spiral Jetty* and the tar seeps side-by-side both as earthworks—one man-made, the other nature-made—suggesting natural agency and articulations beyond words. Both inscriptions beckon beyond human understanding, raising questions about many kinds of marks that we make on this earth. As both have disappeared and reemerged over decades with the rising and receding lake, *Spiral Jetty* has been left dry and inadvertently emerged as a barometer for climate change. As varied meanings seep to the surface, interconnections—between tar and art, sea and sky, weathered rocks and feathered birds—invite revaluations to re-perceive not only an overlooked lake in Utah but also other underappreciated places across the planet, both far afield and right where we are.

Fieldwork requires patience and presence to adapt to circumstances and focus attention on something outside ourselves. The practice forges a suspended sense of time: to contemplate a place through layers of interrelated agency and contingency. To do fieldwork in an unfamiliar site may be easier than in zones close to home, where accumulated habits may warp perceptions. Encountering a place with a beginner's mind, listening to entangled perspectives, complicates our ability to take elsewhere for granted, to deduce a single narrative or otherwise simplify complex environmental and cultural layers—especially as intersecting stories continue to shift.

Life in the Tar Seeps traverses a few of my visits to Rozel Point as if searching through a philosophy of curated stones and decomposing bones. Great Salt Lake's tar seeps and *Spiral Jetty* interconnect through geologic and artistic timescales often present through absence; in kind, much of this book is absent, at times fragmented in form. Like a stratified rock, an excavated sediment sample, or a disarticulating pelican bone fringed with feathers, *Life in the Tar Seeps* is a fossil in the making. It offers a meditation on place. A poetics of space. Great Salt Lake seeps into these pages and between the lines. Like a photograph of Land Art (which cannot be confused with the work itself: to be experienced in person, through shifting perceptions over time), *Life in the Tar Seeps* is a partial representation. Gaps suggest a deeper liquid, viscous story. As tar melts and freezes, my experiences compress a few interpretations of a challenging place, where visitors must constantly renegotiate our relationship with the earth: to watch where we step.

As Great Salt Lake depletes from drought, more tar seeps have emerged along the receding shore. The tar seeps fossilize signs of the lake's abundant life, including pelicans that breed offshore at Gunnison Island. Millions of birds annually migrate to the nearby Bear River Migratory Bird Refuge. Great Salt Lake does not exist in isolation—at the convergence of two of the four major migratory bird flyways of North America. The lake also converges with agricultural runoff, toxic dumps, pollution, and resource extractions affecting those who rely on its essential watershed. A migrating bird that ingests a toxin around Great Salt Lake can die or carry it elsewhere. The lake affects many lives beyond.

With vast ripple effects, Great Salt Lake lives in a massive basin that offers no drainage to gulfs or oceans. The Great Basin drains internally, stretching the Earth's crust across states from California's Sierra Nevada to Utah's Wasatch Mountains, and farther north and south. Larger aspects of this geographic region—of monumental earthworks, archaeological and modernized human settlements, potsherds and rock art, large-scale excavations and extractions, military-industrial installations, and interplanetary simulators—lie beyond these pages in others' books, entangled with possible futures to be collectively imagined.

As a book, *Life in the Tar Seeps* spirals around this depleting body of water to trace an environmental sensibility that continually approaches and distances its interconnected watershed. I think of this exercise-in-perception as a *spiraling ecology:* a way to cohabit our ecosystems at once inwardly and outwardly, microscopically and macroscopically, scientifically and artistically, more integrally than separately. Such an exercise does not require traveling far geographically, if we move beyond our habituated, limited perceptions of ourselves and how we occupy the world. (The word *oecologie* derives from Greek roots in *oikos* [οἶκος], akin to "home," "household," "dwelling place," and interrelationships between organisms and our environments; in kind, our perceptions can be exercised wherever we are.) By deepening our attentions to a particular place over time, disorienting and reorienting ourselves to collective changes within a living ecosystem, we might retune our breaths to the larger pulse of the planet. The practice can quicken or quiet: moment by moment, season by season, year by year, action by interaction. Perceptions shift as a spiral is a shape of expansion, seen as unseen, from a DNA helix to the galaxy of the Milky Way, from a funneling cyclone to water swirling down a drain. This book adopts the furling shape of a *spiraling ecology* to invite a meditation on time in space, unfurling from Great Salt Lake to where you find yourself now.

By the time this book is printed and held in your hands, it will be decomposing. A book is a material object and also a measure of time. Shaped by the sparsity of the high desert, this book holds my photographs that retreat as the narrative melts into other matters: from printed to virtual pages (through QR "quick response" codes, appearing as grids, which can be activated by mobile devices). These augmentations enable more resources to be included beyond print, while standing in for electronic fossils-in-the-making. The book's virtual life will continue to shift through evolving studies and stories around Great Salt Lake that remain in progress, collectively working to sustain the intricate, vulnerable watershed of lives—human and beyond—in the region often referred to as the American West in North America.

As our brains evolve through reading and writing in the digital age, questions evolve. How do we, as a species, co-create a "quick response" to the larger questions outside this book, namely the climate crisis? Can we cultivate care for challenging places that may be overlooked in our midst, here and elsewhere, even those that may never be encountered in person? How do we confront our mortality and vulnerability to recognize kindred dynamics in our living planet? Might we move beyond narrow conceptions of life and death, health and injury, ugliness and beauty, elsewhere and here (wherever "here" may be) to enliven communities of care for entangled ecosystems that evolve over time? Digital visual literacy makes humans less dependent on multisensory knowledges—sound, smell, taste, touch—that help our species interact with the world beyond words. Might integrating diverse and biodiverse knowledges awaken dormant or untapped registers of our bodies, attuning more to bodies of water and land, potentially learning renewable ways to heal a wider world?

There is more to this story, and this book's limitations are evident to this outsider who lived by Great Salt Lake for only a short time. The errors and shortcomings in these pages are my own. Larger studies of Great Salt Lake are ongoing through evolving collaboratives, alongside those who inhabit the Great Basin and the American West, whose stratified yet intersecting stories fold into each other. Continental plates diverge and converge, transforming under our shared atmosphere. This book remains incomplete to reflect what may yet be learned by paying closer attention to wherever we are—as past and future lives entangle, palpable in the present—not isolated but interconnecting the larger life of the Earth.

Writing this preface now in early 2020, back in Washington, D.C., a few years after tracking a motley crew of scientists through the following pages, after writing this book—my perception has shifted to see this "dead sea" with its "death traps" as deeply alive. I no longer view Great Salt Lake through human eyes only, but also try to imagine it through pelican eyes: where migratory bird flyways converge. My attempt to value the lake by following some of its many birds (owls, gulls, pelicans) through photographed encounters falls short of trying to "see" a garden through butterfly eyes, or a mountain through the eyes of a wolf—an attempt that is always inadequate, does not happen overnight, nor does it finish with a book whose end doubles as a beginning. Only after decades did Aldo Leopold reconcile his experiences as a young man into an interconnected "land ethic," writing in his memoir of *A Sand County Almanac* (posthumously published in 1949):

> We reached the old wolf in time to watch a fierce green fire dying in her eyes. I realized then, and have known ever since, that there was something new to me in those eyes—something known only to her and to the mountain. I was young then, and full of trigger-itch . . . But after seeing the green fire die, I sensed that neither the wolf nor the mountain agreed with such a view.

Beyond human perspectives, there is more to be learned from other species as kin, as Potawatomi scientist and storyteller Robin Wall Kimmerer invites (2015):

> In indigenous ways of knowing, other species are recognized not only as persons, but also as teachers who can inspire how we might live. We can learn a new solar economy from plants, medicines from mycelia, and architecture from the ants. By learning from other species, we might even learn humility.

If humans could perceive Great Salt Lake through a pelican's eyes, through a migratory bird's eye view, we might better sense our interdependence with this living, mortal, moving planet.

PROLOGUE: Decomposition

The Microscope being taken into use, this Shy colour'd Butterfly, looks exactly of the same colour as the Thatch which covers Roofs of Houses, the Feathers on its Sides being so beautifull & placed . . . and of such dazzling colours as deserve to be diligently contemplated, it not being Possible to describe them by Pen or Pencil.

—"Commentary," in Maria Sibylla Merian, *Dissertatio de generatione et metamorphosibus insectorum Surinamensium* (1719 ed., John Carter Brown Library)

In my photo stream, there is a gap starting in April 2016. Before the date I was hit by a car: details of rare books at the John Carter Brown Library in Providence, Rhode Island. Etched caterpillars and cocoons. A crested heron. A red-winged cotinga. Flying squirrel. Magnolia blooms. Sprouting leaves. Fish swimming in margins, across maps. My snapshots of aloe greens and sky blues colored black and white images (ship inventories, abolitionist treatises, sheet music) as I researched an opera libretto. Although days focused on unnatural subjects, the natural world seeded my hours.

On the final day of my research fellowship, I had planned to immerse in natural history (including a chrysalis-filled treatise by Maria Sibylla Merian, where each page metamorphosed with butterflies and moths) before anticipated reunions, before my husband arrived from Washington, D.C., for a weekend rendezvous. The day would start with breakfast to celebrate at my favorite coffee shop. A few blocks from my temporary residence, a traffic light signaled to walk. I stepped into the crosswalk.

Beyond pain lay blank darkness. Bright sunlight. A stranger cradled my head in his lap. Lying on the street, I heard his voice: a mouth on a face floating in the sun. *Who? What?* Air ricocheted in my head like a throbbing heart. I couldn't move but had been going—*Where?* Blurring. Blankness. *Don't move,* he said. *There's a lot of blood.* Confusion spun into panic I couldn't articulate. No pain yet: too present it was absent. *I have to go,* I thought. Bright penumbra. Light. Sky. His head floated. Chilly. *No. I have to go,* I thought, *now.* My body wasn't floating, sinking as stone. *Let me go,* I thought urgently, unmoving. Bruised, swelling, bleeding. *This can't be:* slow awaking, deep aching, as if my body were being pulled apart in slow motion. *Your spine may be broken,* he said. *An ambulance is on its way. You were hit by a car. Don't move.* Time reeled forward, back, stood still.

He started to relay an impossible story, a parallel world, blazing as the sun around his face, dark stars without light, constellating on this street that a few days earlier had been cloaked in snow. He had seen me. Hit. Now: everything pulsed, erasing the line between before, after, here. Car. There had been no car. A crosswalk. Pain. *Do you know your name?* he asks. *My name?* I say my name, not certain it belongs here, far away yet close enough that his question makes me wonder if what he says is true. My age and date blur. He asks me the president, a timeline, to place myself in the history of now. I say *Obama,* seasons before the 2016 election, not knowing the question is apocryphal, that the world will collide again, again, again.

Later, after my body realizes that it can barely move, after the ambulance and emergency room and ICU, after the world spirals to blackness and back, after a doctor staples my head together in nine places and wraps my aching flesh and bones, reality sinks in.

My husband arrives from a panicked eight-hour drive. Overnight in a hospital bed, my arm is hooked to an IV, my legs in compressors. Vertigo. Challenging ride home to recover. Vertigo. Days and weeks in slow motion. Vertigo. Greening trees outside our bedroom window. A walker. One step, two step. Again.

Time recesses, so the accident feels like a bad dream. The world around me glows. Life feels suspended over the following year in a new dimension of time.

"[T]his significance is steeped in events, both human and geologic, that happened in the past. But these same sites also exist in the present. The challenge is to find a structure for expressing this deep past . . . and, at the same time, to emphasize not just what is lost to the past and to the site, but also how this loss is apprehended and can produce something new: what remains."

—Ann Reynolds, "The Problem of Return" on *Spiral Jetty* (2009)

Looking through my photo stream from before the accident, scrolling past browned pages of rare books, I spy shells on a bare beach and in glass cases in a campus nature lab. A fan of ruby feathers fills a frame. A flock of taxidermied birds perches under antler racks. Boxed eggs bear handwritten labels: herring gull, scarlet tanager, quail. A snapshot of my laptop shows a live camera on eaglets in a nest at the National Arboretum. I scroll back through the days in reverse. Among more rare books, a bird's eye view from a historic church belfry. A harp on an old headstone. A low wall of stones characteristic of New England. Then, as if out of nowhere, an open book appears with a now-familiar site out of place in Rhode Island—a photograph within a photograph—of four concrete cylinders in the desert arranged to align with sunlight and solstices, patterned with holes to project constellations, large enough to stand upright and walk inside. The photos show Nancy Holt's massive mid-1970s Land Art in remote Utah called the *Sun Tunnels.*

"The panoramic view of the landscape is too overwhelming to take in without visual reference points . . . Through the tunnels, parts of the landscape are framed and come into focus . . . extend[ing] the viewer visually into the landscape, opening up the perceived space. But once inside the tunnels, the work encloses—surrounds—and there is a framing of the landscape through the ends of the tunnels and through the holes . . . Day is turned into night, and an inversion of the sky takes place: stars are cast down to earth."

—Nancy Holt, "Sun Tunnels" (1977)

A year and a half after the accident, in October 2017, I travel to Utah to meet my new colleagues as a visiting professor of environmental humanities. The first place that my husband and I visit is Robert Smithson's *Spiral Jetty,* easier to access than Holt's *Sun Tunnels* (about the same age as me, b. 1975). The site is otherworldly yet smaller than expected. We follow the artwork and stay inside its spiral. We don't roam the mudflats or notice pools of seeping raw oil. Even if we had seen the tar seeps, we wouldn't have known to call them by that name—they would have seemed negligible smudges near rotting pylons and abandoned drilling equipment. Our focus turns to the sun-spattered spiral of salt-encrusted basalt. A crisp, sunny, clear autumn morning. The lake spreads blue toward purpled islands to the seeming ends of the earth. A scenic route takes us back to the city, up creeks and canyons that slice foothills, fringed with flaming leaves under snowcapped peaks.

Come January, the night before teaching my first class of that semester, my father-in-law dies of pneumonia and complications from Parkinson's disease. As I fly in and out of the Great Salt Lake basin for his funeral in the Midwest, Utah's polluted inversions seep into my lungs and make my own breathing difficult. Through those blurry weeks of early 2018, a quiet solitude polishes my lens to see Great Salt Lake as something more than a wintry backdrop. Come February, I make my first visit to the tar seeps. As an outsider, not a scientist but rather a curious and searching human, living apart from family and friends in a city of relative strangers, I start to appreciate how the lake gives life not only to the tar seeps and *Spiral Jetty* but to the entire Basin and Range, even beyond.

"This is the heartbeat. It's the place of transformation. The arteries that go into the lake look like what's going into your body. We have to honor what people were doing hundreds of years ago. Migrations here included the language of 'well-watered and eternal waters.' We still tell these narratives of ourselves—those narratives stick with us. But when there aren't eternal snows on mountains, it's hard for us to accept that change."

—Jaimi Butler, Great Salt Lake Institute, in conversation (2018)

After returning from Utah to Washington, D.C., I scroll through a photo stream that I share with my husband. Many of my photographs come from the West: swirling sandstone canyons, granite mountain peaks, the lake's orange halophiles and black tar seeps of natural asphalt. From a chair in Washington, D.C., I glimpse other places—around Utah, California, Montana, Washington, Switzerland, Rhode Island, more—passing in front of me on a scrollable, virtual contact sheet on a laptop. The act of swiping bypasses days, months, years: virtual time travel, reduced to snapshots. I swipe forward and backward.

Then, a photo stops me still: taken with my husband's phone, the date of my accident. It frames a body in a hospital bed, head cropped at an inhuman angle, so you cannot see my face: just my matted hair, arm in an IV, and a sheet wet with blood. I have not seen this photograph before—feeling as if my body is back in the crosswalk, lying on manmade asphalt, gazing at the sun around a floating face, like double vision, looking at myself through someone else's eyes.

Through that photograph, I perceive the shortcomings of any attempt to view a situation through someone else's eyes. As if looking at a pelican or gull in a tar seep, I am daily reminded of my own shortcomings, vulnerabilities, and points of getting stuck. Any human gets stuck again and again on this planet. Random accidents in our lives make us take stock and recalibrate. Our perspectives are partial, incomplete without each other. Yet the incompleteness—the gaps, signals of disconnections—may enable potential new, even renewable, connections. The things that are hardest to understand, or easiest to overlook, become places to pay more attention: to consider what lies unnoticed, right where we are, to make our next steps.

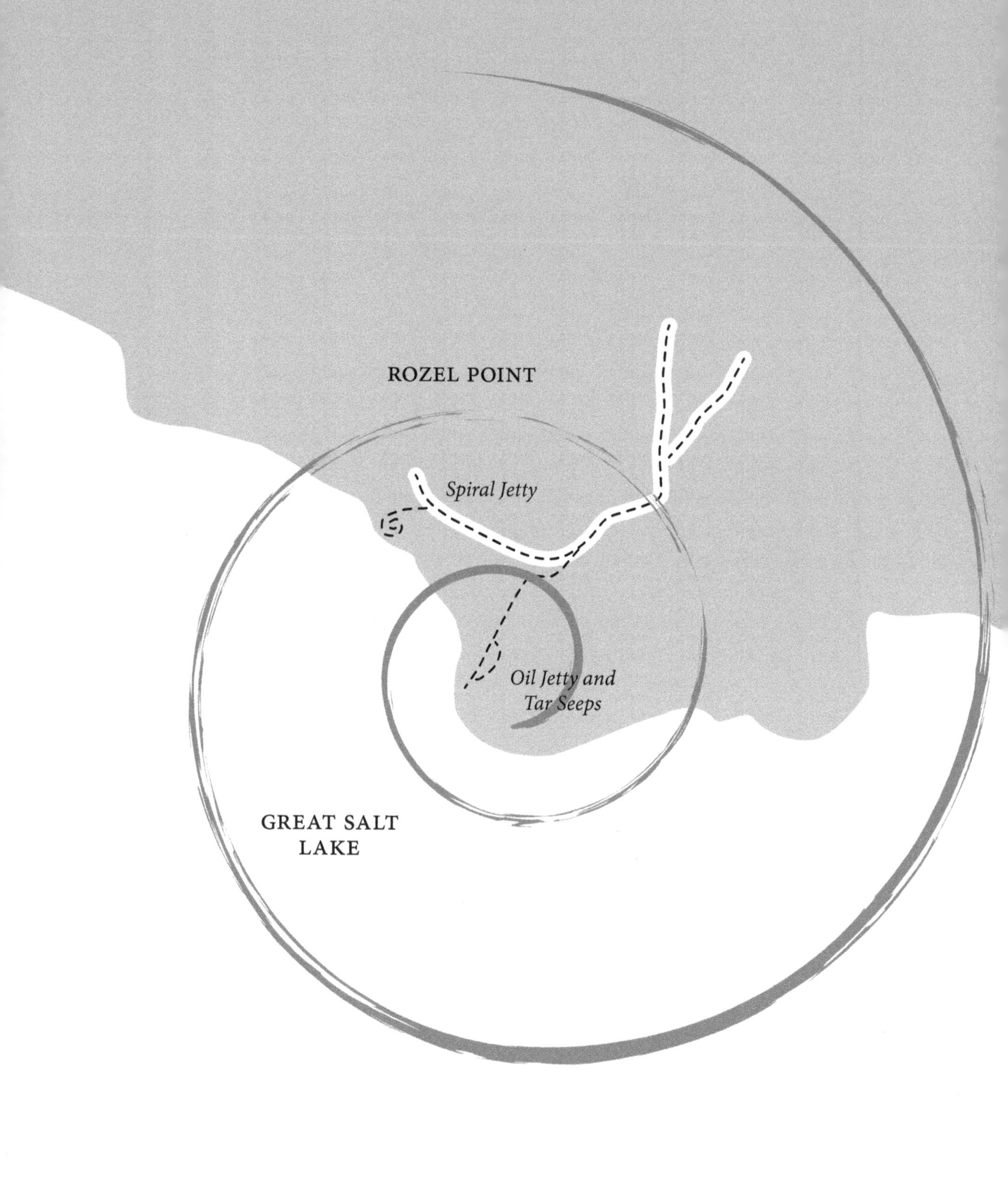
ROZEL POINT
Spiral Jetty
Oil Jetty and
Tar Seeps
GREAT SALT
LAKE

GREAT SALT LAKE

I.

the birds . . . remind me of what I love rather
than what I fear . . . They teach me how to listen.
— Terry Tempest Williams, *Refuge*

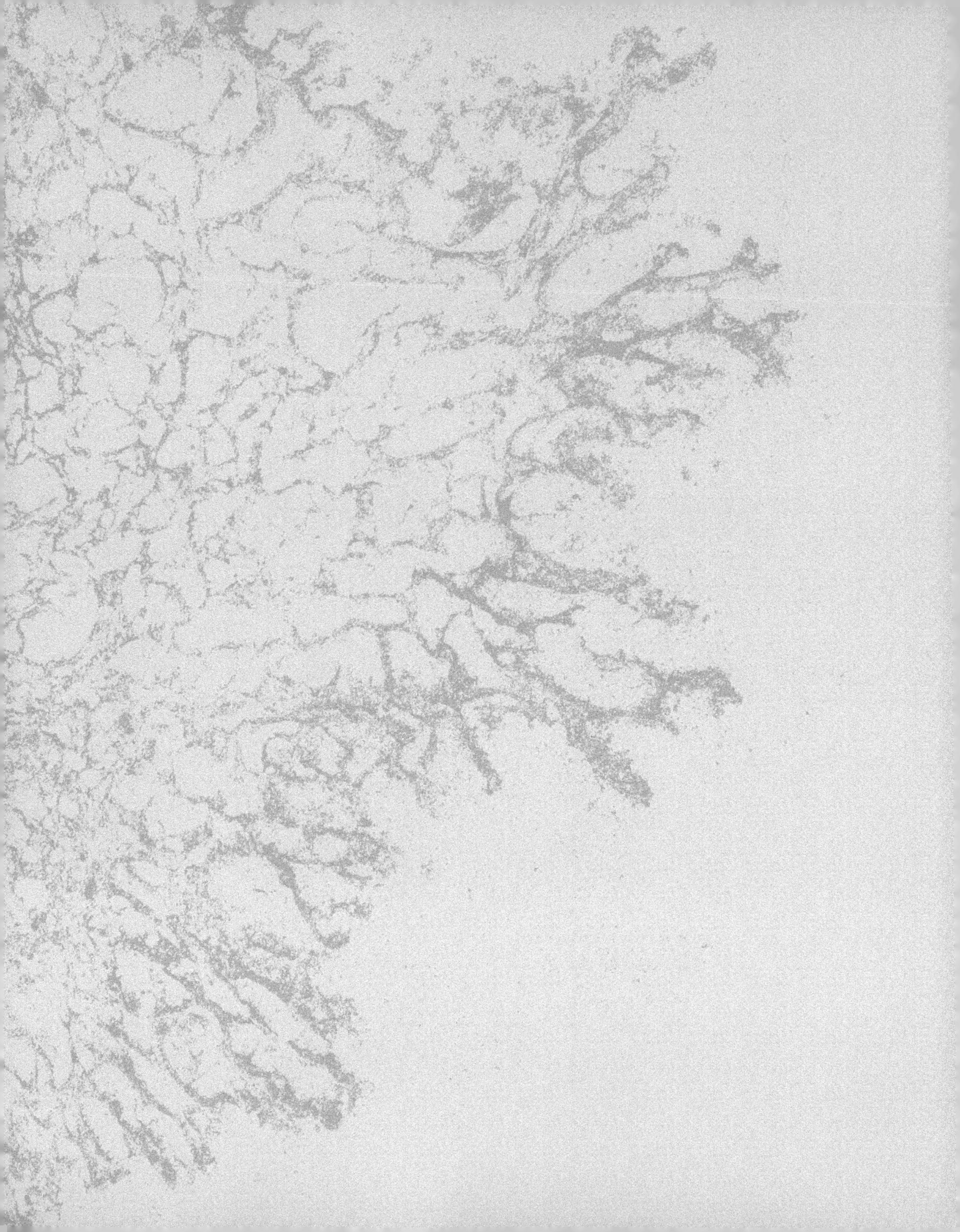

Death Traps

In the rearview mirror, the Wasatch Mountains of Utah rise from the Great Basin. Low hills shoulder limestone caves, tucked into parched slopes of tall grass that roll toward Great Salt Lake like ancient waves. Long-gone shorelines band the hills like rings in a bathtub. A two-lane paved road cuts between dips in the knolls, edged by marshlands that spread through the Bear River Migratory Bird Refuge, one of the largest migratory refuges in North America. On this February day, the chill has fogged into crystalline snow, blurring the marshlands, hills, and mountains until we lose sight of all distances, just the road around us. Almost imperceptibly, the air smells like rotten eggs.

"You always hear that the lake is dead, but it's so alive that it smells," says Jaimi Butler.

Jaimi is the coordinator of the ten-year-old Great Salt Lake Institute (GSLI), an interdisciplinary environmental research center dedicated to this understudied ecosystem. She is driving a blue minivan with a fiery flame decal on its door. Her five-year-old daughter, Cora, sits in a car seat behind me watching *The Angry Birds Movie* on an iPad, beside Greg McDonald, the regional paleontologist for the Bureau of Land Management. We are heading to Rozel Point, on the remote north arm of Great Salt Lake.

We left Salt Lake City after dawn and drove north, skirting the east edge of the lake. Greg and Jaimi are meeting for the first time, hoping to partner on a project to set up camera traps on the lake's tar seeps—also known as oil or petroleum seeps, and nicknamed "death traps"—which lie near the iconic Land Art work by Robert Smithson, *Spiral Jetty.* Jaimi has lived in Utah her entire life, over four decades; Greg has lived here for two years. As a visiting professor in environmental humanities at the University of Utah, I am a newcomer to this place and have lived here barely two months. A colleague knew of my interest in *Spiral Jetty* and connected me with Jaimi, which is how I've found myself on Jaimi and Greg's quest: tracking what animals get caught in the tar seeps, glimpsing fossils in the making.

"We haven't documented animals getting trapped in these seeps," Greg says. His career has led him to visit tar seeps and tar pits across the Americas, with a focus on ancient sloths. "Not all tar pits are the same," he says, referring to the famous La Brea Tar Pits. "In the Los Angeles basin, sedimentation covers the tar pits, then an earthquake or other fracture will make it seep to the surface, and it spreads out like flypaper. Utah also has earthquakes, so at Rozel Point we may see similar processes—but these seeps are thin, not deep pits like La Brea, and may change when the lake rises and falls."

I know little about Great Salt Lake, and nothing of tar seeps. A decade ago, purely by accident, my husband and I walked past the La Brea Tar Pits on our way to the Los Angeles County Museum of Art. The strangeness of the urban tar pits surprised us enough to stop, briefly, to peer at the ancient animals in sticky asphalt. As a kid in San Francisco, I wasn't interested in dinosaurs or fossilized bones. My family's summer road trips drove right past Great Salt Lake without stopping, en route to national parks.

I have taken this drive only once, back in October with my husband, coming cross-country from Washington, D.C. Our destination was *Spiral Jetty.* I had dreamed of following its massive spiral of salt-encrusted, black basalt, unfurling three times counterclockwise into the lake. I am interested in how natural spaces—

national parks, wildlife refuges, nature preserves, Land Art—require collaborative stewardship by land managers, scientists, artists, curators, Indigenous communities, and other partners who care for living environments as they evolve over time. As different groups share their stories of a single place, they can merge approaches and resources to support each other's coexisting narratives.

"We would be underwater here," Jaimi says, steering the minivan along the highway. She nods toward parallel ridges in the hills, called benches, that mark the ancient lake levels. She has spent her adult life working at Great Salt Lake, first studying harvests of brine shrimp and diets of eared grebes and, now for the GSLI, coordinating partnerships around its shore. The GSLI, based at Westminster College in Salt Lake City, is the local steward of *Spiral Jetty*, and is joined in its efforts by the Utah Museum of Fine Arts, the Utah Department of Natural Resources, and the New York–based Dia Art Foundation. Jaimi travels monthly to the artwork—almost three hours each way, over remote dirt roads—to take water and salt samples, look for tagged pelicans, host local science teachers, and monitor the number of visitors to the artwork by checking a road counter. "*Spiral Jetty* was underwater for many years," she says. "Soon after Smithson built it, the lake covered it up. It didn't reemerge until 2002."

As we wind past low hills, Jaimi and Greg name the ancient lake levels. Each shoreline is a ghost of a freshwater sea, etched as a horizontal line across the slopes, where the lake levels remained long enough to leave marks. Stansbury, Bonneville, Provo, and Gilbert: the state traces its name to the indigenous Ute Indians, but the basin's shorelines are now named for explorers who were not native to the region.

The hills are muted on this gray day. Many afternoons when I leave my office on the sloping campus of the University of Utah, the lake's golden sheen is hypnotic, an ethereal sight that stops me in my tracks. Yet when I talk with locals, Great Salt Lake is often described in pejorative terms, as stinky or ugly.

The first known written account of the "inland sea" came in 1776 from a Spanish missionary expedition that never reached the lake but heard of it from the Timpangotzis Indians. Over the centuries, freshwater wetlands sustained ample plant and animal life, and archaeological evidence in shoreline caves suggests how Paleo-Indians foraged, hunted, and farmed in communities that fluctuated with lake levels. Floods and droughts influenced Fremont and, later, Numic hunter-gatherers who became the ancestors of the Ute, Shoshone, Goshute, and Paiute Indians. As settler-colonists arrived—explorers, trappers, and miners, along with Mormons who considered the valley their promised land—most tended to overlook the lake's life. In 1849, Captain Howard Stansbury set out to circumnavigate the lake and said it carried the "stillness of the grave" with "bleak and naked shores," where almost "nothing is heard . . . unless it be that very rarely a solitary gull." His references noted "bitumen" resembling "seaweed of the ocean," "softened by the sun, and completely frosted," and "black mud . . . impregnated with all the villainous smells which nature's laboratory was capable of producing." Many accounts of the lake carried the connotation of death, which explains why the lake has been understudied. For a long time, it was considered a dead sea—almost as salty as the Dead Sea—and that reputation has overshadowed the lake's vitality. For many people over the past century and a half, Great Salt Lake has been virtually hiding in plain sight.

By now it's midmorning, but we're still some distance from Rozel Point. "I was out there in October and found four barn owls in the same tar seep," Greg says. He explains how the raw oil of the seeps emerges from tectonic fractures and creeps across the lake's mudflats. "The owls probably got stuck in summer, maybe chasing mice. There were also pelicans in the tar volcanoes."

The marshlands retreat behind us. The Bear River Migratory Bird Refuge covers eighty thousand acres of mudflats, marshes, uplands, and open water supporting the annual migration of over 250 bird species: white pelicans, eared grebes, American avocets, tundra swans, among many others.

As we speed down the highway, a blue billboard reads "SAFETY FIRST—ALWAYS." Behind fences, dirt roads crisscross the hills where cows graze. A smaller sign marks "Orbital ATK Propulsion Systems." Orbital makes rocket motors and boosters for NASA, and the dirt roads are fire breaks in case of explosions. A small brown sign points to the Golden Spike National Historic Site. We turn.

I remember, weeks earlier, seeing a map in the university's special collections showing a massive blank marked as The Great Basin. The map was part of the report of John Charles Fremont's 1843–44 expedition that fed the American colonial settlement of the West. An arced line of text mimics a topography "surrounded

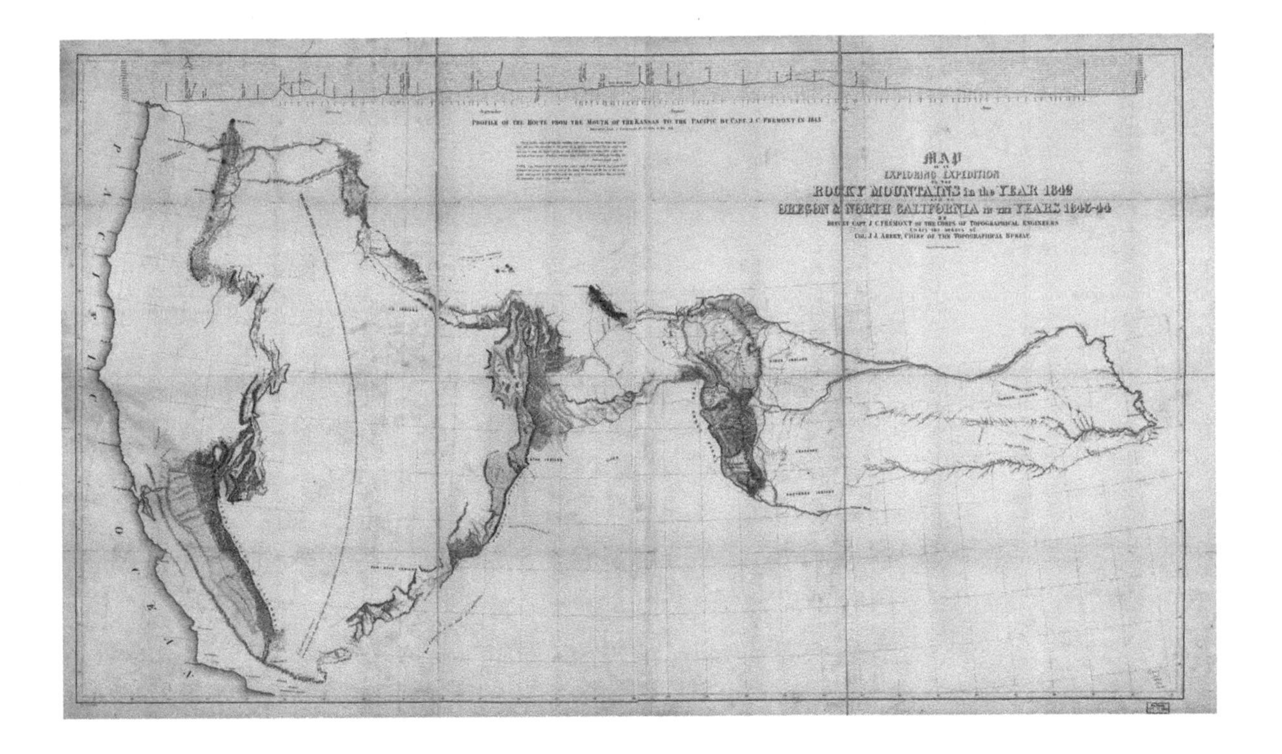

by lofty mountains: contents almost unknown, but believed to be filled with rivers and lakes which have no communication with the sea." The lyrical negation, spelled out over a massive blank spot on a detailed map, suggested the wide berth of the basin and its overlooked histories. An earlier map showed the *Valle Salado,* or salty valley, whose "Western limits of this Lake are unknown" with a river to "probably be the communication between the Atlantic and Pacific." Early trappers mistook the lake for the Pacific Ocean. It was recorded that the Western Goshute tribe referred to the lake as *Pi'a-pa,* meaning "big water," or *Ti'tsa-pa,* meaning "bad water." Different Indigenous groups had used the lake as a trading hotspot, to get salt and to visit hot springs for healing, but the lake became an edge in the newfound promised land. After 1847, Mormon settlers laid their signature grid system for Salt Lake City and carved the southeast shores into homesteads. Around farms grew towns, mining industries, entertainments like Saltair and the Bonneville Speedway, and military testing and training ranges. As later maps filled with new details, overwriting Native place names and communities, the north and west shores of the lake remained remote, accessible only by a few rutted roads and flat-bottomed boats. The road on which we're traveling today is one of the few access points to the lake's north arm.

The pavement ends at the Golden Spike National Historic Site in Promontory. We stop briefly at the visitor center, the last available public restroom. We are the only car in the lot. A plaque on a commemorative obelisk describes how, in 1869, a golden spike was driven to complete the first Transcontinental Railroad: IT ACHIEVED THE GREAT POLITICAL OBJECTIVE OF BINDING TOGETHER BY IRON BONDS THE EXTREMITIES OF CONTINENTAL UNITED STATES, A RAIL LINK FROM OCEAN TO OCEAN. The railroad link came shortly after the United States Army slaughtered hundreds of Shoshone in the nearby Bear River Massacre. Other local Indigenous communities were displaced by force and devastated by disease. The railroad was built largely by immigrant laborers, predominantly from China, and many of them died in the course of construction. Now isolated in winter, Promontory seems inhabited mainly by ghosts.

When my husband and I visited *Spiral Jetty* the previous October, we brought a pamphlet from the Utah Museum of Fine Arts that had information not only on the artwork but also on basalt rocks, microbial halophiles, and oolitic sands, with driving directions in ten bullet points. Number five read: "Cross a cattle guard. Call this cattle guard #1. Including this one, you cross four cattle guards before you reach Rozel Point and *Spiral Jetty*." The instructions recommend packing plenty of food and water, wearing weather-appropriate clothing and waterproof boots. There's no cell service, so if your car breaks down, you may be literally up a dry creek.

Back on the gravel road, Jaimi's minivan bumps against ruts. The fog seems denser, and occasional flakes of snow smear the cracked windshield. As we descend Promontory Pass, low hills rise from the gully that cuts through ranchland. It is easy to imagine the landscape's contours shaped by water.

The road forks. Jaimi knows where to turn and where to cross the cattle grates. Cows amble across the rough road and graze in sage. Fences stitch the fields. We pass one pickup truck and see fewer and fewer birds.

The snowflakes stop. Slowly the fog starts to lift. Great Salt Lake comes into view, a silvery sweep in the distance. Sun burns through clouds. The glow shifts. Rock islands appear and disappear as elusive mirages. The landscape appears otherworldly, almost extraterrestrial.

Usually when I see the valley, walking downslope from campus to my apartment, the lake spreads from the base of the snowcapped Oquirrh Mountains, silhouetting at sunset and purpling into starry dark. Here, away from the city's skyline, the mountains sink like teeth of an open jaw on the verge of swallowing the sky.

With each turn of our drive, I feel that we are going backward in time. Years fold in on themselves like geological strata. Time no longer feels linear. As the fog lifts from the circling mountains, dark ridges appear as supine sleeping giants around a frozen cauldron or an extinguished campfire. It is easy to imagine these hills as animate, mythically arising from primordial shores.

The road splits to two dead ends. The near end of the fork leads to the tar seeps; the farther to *Spiral Jetty.* We park, stretch our legs, and scramble down the rocks onto cracked mudflats that seem to spread for miles toward the lake. Rusting drums and forgotten piers mark abandoned attempts at oil drilling. Apart from the gravel road and rotting wooden pylons, the landscape appears posthuman.

As we cross the mudflats, heading for the seeps, my face chafes from the cold and my nose begins to run. Greg doesn't seem to mind the chill, toting a camera bag with bare hands and walking alertly. Jaimi and Cora follow. It seems possible to walk far in any direction.

At a boulder that resembles bleached coral, Jaimi and Greg kneel. It holds water like a religious font. The limestone appears dead but brims with ancient marine life.

Jaimi's face twists with concern. She points to tire tracks where an ATV crisscrossed the mound. The tracks are fresh. "They aren't supposed to be here," she says. "But who's going to stop them?"

We start walking again. Winds whip us with salty air. Step by step, I am hyperaware of my body, my mobility, my clothing: goose-down jacket, turtleneck sweater, snow pants, long underwear, tennis shoes, wool scarf and mittens, in which I wiggle my fingers so they don't go numb. Under a thick hat, my hair covers nine scars where my head was stapled together, almost two years ago, after I was hit by a car in a crosswalk. The spots throb along my scalp when I am tired or overworked, sometimes a bellwether for incoming storms. Today I barely feel them walking this barren ground, engulfed in chilly wind, marveling at wonders and breathing deeply. As I gaze over the sweep of sand and mud, black mounds start to

constellate out of the seeming voids: dense starry clusters, not light but black, not in the sky but on shore.

"Once you learn to identify a seep, you start to see them everywhere," Greg says.

The tar seeps spread across the sand like frozen black puddles. Some are flat and thin; others are bubbly and raised: tar volcanoes. Most have liquid edges, as if they're melting. One tar volcano glistens wetly, but when Greg pokes at it, he hits a solid surface, firm as hardened plastic. The edges of the seeps spread in delicate laceworks where their heated ooze met the mud and radiated in fractal-like patterns. Up close, they are riddled with disarticulating feathers and bones. Their shapes suggest inversions, specific to this place yet surreal enough to spark comparisons. They remind me of blackened versions of rust-colored canyons farther south, where dunes and drifts compressed over millennia into swirled sandstone. Utah is full of stony seas and reefs with fossilized barnacles, shells, and bones. The seeps seem to emerge by negation, like an etching in relief. The not-here comes to mind as we observe what is here.

"We couldn't do this in summer," Greg says. Warm weather wakes up the seeps, where a step can get you stuck. A dog almost died in the seeps, saved by strangers who wrangled plastic bags and car mats around its sinking body until it was dislodged, coated in 40 pounds of tar.

"In Los Angeles, you can't do this," Greg adds, referring back to the La Brea Tar Pits. "It's too developed and fenced in. Here we've got a fairly raw site that has had minimal disturbance. Its natural events are happening with minimal human activity, so it becomes an important baseline for understanding the whole process. We can't go back and photograph a mammoth getting stuck or a saber-toothed cat getting stuck, but we can certainly look at the conditions and see what's getting preserved."

We keep walking through the chill, pulling our coats tighter. More tar seeps appear, pocking the mudflats. Our shoes accrue damp sand but, thankfully, not tar.

"Look at that pelican death assemblage," Greg says, stopping and pointing to a group of tar volcanoes.

We walk toward the archipelago of melted black mounds, skirting the edges to examine heaps of tar and stuck bones. Tar is considered a perfect preservative—

utah
0E5

freezing, drying, encasing organisms as when they were alive—fossilizing a life span in fragments.

"There's a skull," Greg says.

"And a bill," Jaimi adds, pointing to the signature hook in the beak.

"There's a humerus," Greg says, pointing to another bone and noting the disintegrating collagen.

Jaimi points with excitement into the bone pile, drawing my eye to two bright green, numbered metal tags. These come from the GSLI's pelican count. The Pelican Roundup tags birds, alongside tracking efforts like PELIcams and PeliTracks, both to better understand the birds and to educate the public. The GSLI involves students in these studies, and Jaimi plans to pair a student with Greg to set up camera traps on the seeps in summer. The PeliTracks—GPS transmitters worn by the pelicans—are funded by the Salt Lake City International Airport. Pelicans pose a danger to—and are in danger from—air traffic; a flock of birds could down a plane. An adult pelican's wingspan extends around nine feet.

"White pelicans are understudied," she says, pointing offshore to Gunnison Island, a prime nesting site that provides refuge to about twenty thousand white pelicans annually. "Ironically, the lake is pretty perfect for pelicans, since the railroad causeway cuts off the north and south arms, protecting them from disturbances. Pelicans nest here because there are no predators. They trade food and fresh water for safety. Because they feed at the Bear River Refuge, on their first flight they have to fly over thirty miles right over this shore, so a number of them get trapped here."

Where the Bear River delta meets the lake, the refuge comprises the largest freshwater part of the Great Salt Lake ecosystem. It sits at the convergence of the Pacific and Central Flyways. For the millions of migratory birds that annually come through the refuge, many use the area to nest.

"People don't think the lake is worth anything aesthetically, biologically, or economically," Jaimi told us in the car. "It's hard to study fine-grain questions because it's hard for scientists to get out there. But we want people to steward the lake. It's a place of paradoxes. It doesn't have an icon. Who connects to brine shrimp and microbes? People come expecting to find romantic red rock but don't find it. It's stinky. You go out and find dead birds. You float and itch when you

swim. There are all these funny, weird things about Great Salt Lake that aren't always described."

"People know pelicans," she continued. "They have those enormous bills." Her voice grew animated, describing the birds' bright white bodies with black-tipped underwings. "They're beautiful and charismatic. They're one of the signature birds of the area and may be a way for people to care about the lake. If we could connect people to pelicans, we could connect them to uphill water diversions, climate change, and impacted marshlands."

Now Jaimi picks out the green metal tags, laughing about how she loves to find dead birds.

"Dead is an indication of how much life is here," Greg says. He mentions taphonomy, the study of fossilization—"a subdiscipline in paleontology, like paleo-forensics. It basically asks, 'Why are these dead bodies here?'"

"That's what we're hoping to see from the camera traps," Jaimi says, explaining that the cameras are triggered by movement and heat as animals enter their vicinity.

"And if they aren't caught in the seeps," Greg adds, "they're washed up on shore, slowly rotting with weathering."

"Many birds are pickled by Great Salt Lake," Jaimi says. She packs up the green metal tags in her backpack, describing the annual tagging of baby pelicans on Gunnison Island after they fledge. Since the chicks can't fly, they squawk as banders try to pin their wings. When the event is filmed, the birds' bills clamp around the lenses, threatening to bite. "It looks adorable," Jaimi says, "but at the same time like an alien abduction."

Camera traps get humans out of the way. "Non-invasive wildlife documentation is usually set up on game trails and trees," Greg says as he surveys the tar seeps. "Here, they'll be more exposed."

"I hope no one disturbs the cameras," Jaimi says, reminding us of the ATV tracks. "Maybe we should leave signs that say it's a student project."

I wonder what else might provide a deterrent, what makes a person respect oolitic sands or dead birds, unfamiliar places or strangers.

"The unexpected is part of the biggest thing for us," Jaimi says, explaining that the GSLI collects data on a range of topics but needs more researchers to piece together the lake's puzzle. "We need a holistic approach to see how all animals use Great Salt Lake."

"These bones are still disarticulating," Greg says, pointing at the assemblage stuck in the tar volcanoes. Disarticulation occurs both in language and bones, detaching at joints and breaking into parts. "In tar, sediments will slowly bury the bones, and then they become part of the fossil record." He gestures beyond the tar seeps. "If you have an animal that is used to wandering through the area in cold temperatures when it's not sticky, it sets up a trail." Conditions change with the seasons, warming up the seeps. "Suddenly it's following its trail, and part of it crosses a sticky tar seep—it becomes an entrapment event. We need to be thinking about what the fossil sites are telling us. That becomes part of the story."

The story is unfolding before our eyes, in slow motion, swirling together past and future in the present chill. Beyond citing disarticulations, Greg refers to the tar as natural asphalt. Riveted by the seeps, I barely think of manmade asphalt, of the car hitting me, of my head bouncing off the metal hood and meshing with pavement. I try to imagine how pelicans land in tar seeps; I don't think of my scalp stapled back together, of a walker supporting my steps, of rearticulating language over months, relearning to physically move, to read and write. The tar seeps wouldn't be mistaken for streets, but animals who cross this hot asphalt get stuck and die. Slower than a car crash, the seeps enact a different kind of collision—yet with both you don't realize you're stuck until it's too late.

"Let's find those owls," Greg says.

We follow him across the mudflats, continuing away from *Spiral Jetty.* Giant sloths and mammoths once roamed the ancient shoreline, but Greg hopes to see birds, snakes, and small mammals. His dream is to rent a Giddings core drill and take a core sediment sample, to reassemble species across sediment layers and geologic eras, but the cost is prohibitive. Partnerships like the one unfolding with the GSLI may help researchers' efforts to study the lake.

"You need some gypsum," Jaimi says to me, stopping and dropping to her knees, digging with a stick. Like a pig sniffing truffles, she knows where to poke and prod the mud. She pulls out a crystalline wedge and encourages me to take it, then digs up more wedges for Cora. "When we travel or go to conferences," she says, "we bring gypsum, and people love it."

I turn the crystalline crescent over in my hands. Light seeps through the glassy surface. Jaimi and Cora dig for more gypsum. I can only imagine a childhood of repeated visits to *Spiral Jetty,* how that spiraling shape would whorl and shape you.

"Gypsum, salt, and evidence of water are found on Mars, just like here at *Spiral Jetty,*" Jaimi explained in the car. "As a mineral lake evaporates, both gypsum and salt are left behind. NASA will use gypsum to find salt deposits on Mars that are remnants of mineral lakes. Not many places on Earth are like that. Great Salt Lake is one of them." As she turns over the gypsum in her hand, her earlier question resurfaces: "What if you could also find halophiles: life on Mars?"

Halophiles are salt-loving microorganisms that grow in little water. They tint the lake pink or orange, muted or vibrant shades. The GSLI's Director, Bonnie Baxter, partners with NASA to study the extreme environment of Great Salt Lake as an analogue for Mars. Some of the lifeforms found here are among the oldest on Earth. "In 2020, they're sending a rover to Mars to collect samples," Jaimi continues. "What do you do with those samples? On Mars, they'll put them in a parking spot. Will these things stay alive in jars for twenty to thirty years? We can experiment here and learn how to preserve Martian samples.

"It's more likely that halophiles will be our aliens," she adds, "not green little men."

We pocket the gypsum, and I feel awkward, as if stealing a gem. I'm used to leaving things in national parks and protected areas, but admittedly, if allowed, I like to bring home a rock. A geologist would be confused by the random stones scattered around my house. Some people who visit Uluṟu, or Ayers Rock, in Australia reputedly take stones but later mail them back, as if the stones miss home. I imagine a plane flying overhead and casting infrared light until the whole Salt Lake sparkles with gypsum. Maybe it would reveal a lost Atlantis. Or it might reveal all the toxins that are regularly dumped in the lake. Jaimi calls what is unseen the lake's underbelly. She also refers to it as the lake's memory.

I wonder how the lake remembers: as a *body* of water. A few months ago, when a masseuse kneaded an ache deep in my thigh where the SUV made contact, she said: *The body remembers.* In Arches National Park, several hours southeast of here, signs tell visitors not to cross certain rocks to preserve Indigenous rock art: HEALING IN PROGRESS: PLEASE STAY ON DESIGNATED TRAILS ONLY. Perhaps the lake's memories accumulate to impact its existence over time, the way personal events compress. Even as events fade in our minds, our bodies remember: with an ache or stiffness, sensitivity to noise, a residual imprint. The lake may remember through seepage, erosion, or dispersal. Tectonics suggest that stones remember in their way, storing up tensions until they quake.

Greg's path zigzags, as if he has lost the scent, then found it again. He pulls out his yellow field notebook, turns to an entry from October, and reads the coordinates for the owls. He calibrates his GPS unit. "This way," he says, walking with purpose. Jaimi, Cora, and I follow him over the cracked mudflats.

Following Greg's dark parka, his white hair blowing in the wind, I am reminded of my great-uncle Fritz. He taught me to look for patterns on large and small scales. In the Cascade Mountains of Washington State, over three decades ago, he prepared cellular crystals of household liquids—orange juice, even his urine—on microscope slides to distill patterns. He photographed and videoed them, often accompanied by monumental music like Handel's *Messiah,* waiting for my response (*ooh!* and *ah!*) before revealing the cellular source (*ew!*). Then he'd take me outside to his telescope that magnified similar patterns among stars. Before retiring to the Cascades, he had lived across the United States, the South Pacific, Central America, and Africa, working in hospitals on what was called human ecology. I remember him following birds: going out of his way to visit estuaries, carving duck decoys, watching ospreys swoop down the river. His nickname for me was La Paloma. After his death a quarter century ago, I half-seriously wondered whether he could transform into a bird. While my growing fascination with the tar seeps now surprises me, it is a reminder that in the company of someone who cares, you start to care too.

We walk and walk through the chilly air. You could get lost here if you don't pay attention. Most days in the high desert of Salt Lake City (elevation over 4,200 feet), my body dries out, bringing a bloody nose, or my breathing is stifled by smog. A seasonal inversion—a layer of trapped pollution—hovers like a toxic lid over the valley, aggravating health conditions like asthma, increasing the rate of strokes and heart attacks, and affecting pregnancies. People wear masks on bad days, in an attempt to keep toxins out of their bloodstreams. (There's even an app for that: UtahAir.) Decades after the nuclear fallout from the Nevada Test Site, people come from near and far to ski, climb, and hike. Tech companies are transforming "Small

Lake City" into a satellite of Silicon Valley. Dangers hide in plain sight, like the scarred slope of Rio Tinto Kennecott—an open-pit copper mine that collapsed in one of the largest recorded landslides, with a human-triggered earthquake—large enough to be visible from outer space. Carbon emissions grow with industries, as the population swells. In my short time in Utah, I have lain awake at night during inversions, coughing and feeling choked, hyperaware of the basic necessity of breath. Jaimi joked earlier: "What happens in the basin stays in the basin." The same might be said of Earth: What happens on this planet stays on this planet.

Underfoot, the sand is firm and solid, but it's not too hard to imagine the shifting lake, rising and receding, attracting different animals to the edge. The ancient Lake Bonneville spread a quarter of the size of Utah and lost much of its fresh water in the biggest flash flood in history, 14,500 years ago. Fast-forward five hundred years: the lake almost disappeared from drought. Fast-forward again: megadrought is expected to hit the Southwest by 2050. I try to imagine the flats in summer with the stench of sticky tar, beating sun, and mirages on sizzling horizon, dust on our skin and in our lungs. If megadrought hits, the environmental shift may unleash dust storms, buried toxins, dormant viruses, who knows what else. Smithson once described how this landscape "whirled into an indeterminate state, where solid and liquid lost themselves in each other. It was as if the mainland oscillated with waves and pulsations . . . The shore of the lake became the edge of the sun, a boiling curve."

The edge of any promised land emits an apocalyptic aura, as if setting the stage for future salvation or the end of the world. Trying to find alternative narratives for climate change—beyond apocalypse, prophecy, elegy, or tug-of-warring tropes between progress and loss—can cause us to chase the tails of our tales. Shapes of stories recur to mark the edges of our fears, so our tellings fall into predictable patterns and fossilize, separating us from the animals that we are. All of us are gloriously, yet vulnerably, entangled—but so many stories arise from fear of death rather than awe of life, disengaging the individual from the communal, and the human from the nonhuman. The climate crisis isn't a linear narrative; it's more like tar seeps, where a step can get us stuck.

"Here they are," Greg says, finally, and stops. He points to a sprawl of seeps where the dead owls lie.

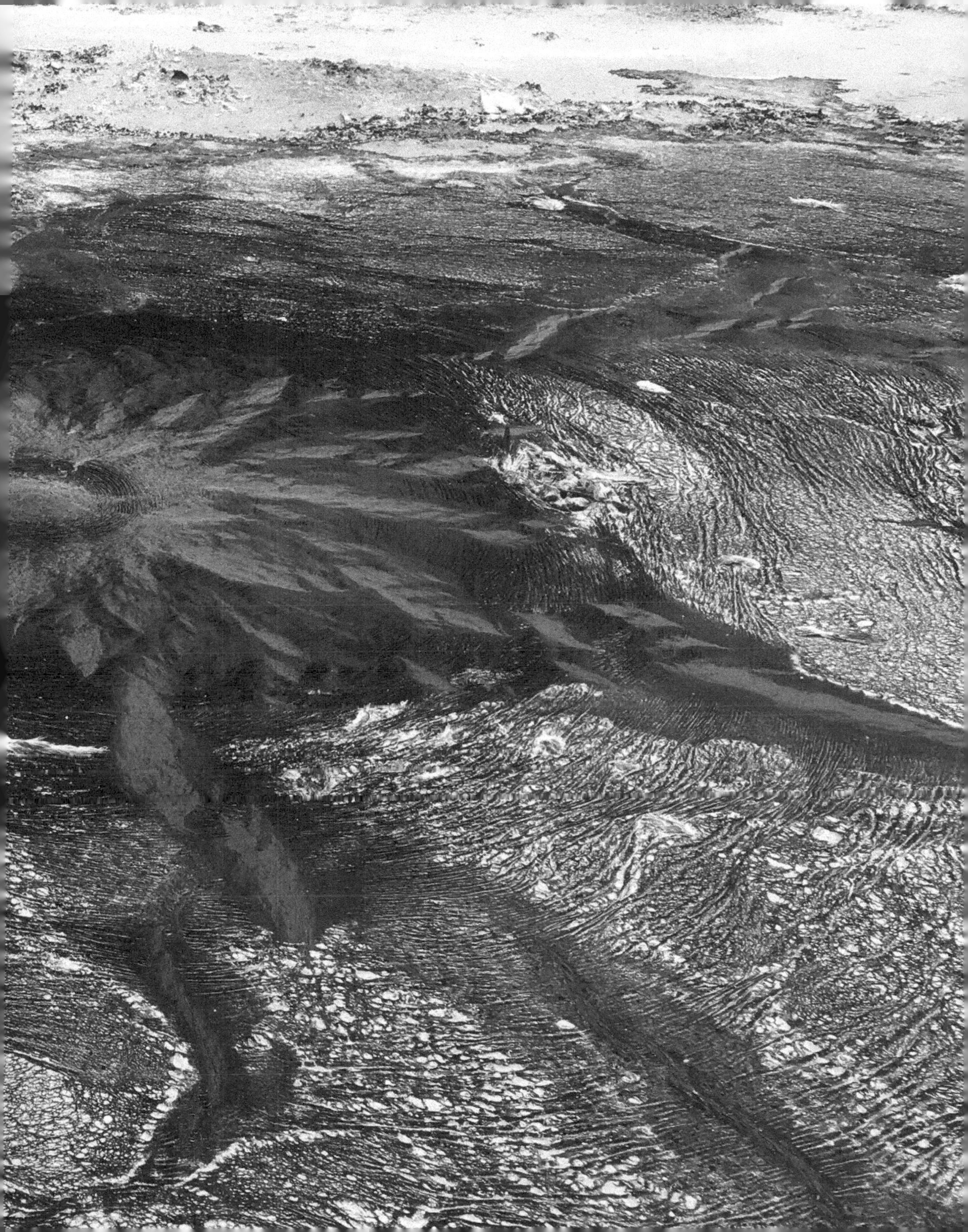

Bones and feathers splay in disintegrating states. The seeps pool together as a shallow black pond, pocked with small tar volcanoes. One volcano is larger than the rest and radiates like a big black star. Its tentacles swirl into what seems an aerial view of a burnt alluvial plain. The wind feels like a giant's breath, not heated but chilling our bones.

The black star seems alive, despite its hibernating state. I follow its whorled tendrils and see an octopus. Then, an anemone. In geologic slow motion, the owl bones and feathers pull apart within the previously melted, now frozen, tar. Melt, freeze. Melt, freeze. Melt, freeze. The process repeats itself with cycling seasons.

Close enough to touch the bones, we keep some distance. We circle the edges of the seeps. Riveted by the star, I frame photographs with my phone. My mind recedes into my skin, gravity, and gait. Pulse by pulse. Breath by breath, I regress from thought to instinct to wonder. My eyes, ears, and flesh coordinate as a body that spent centuries evolving to move. My hands use the camera not as a tool to survive but to frame my encounter: focusing, zooming, distilling the multisensory moment into a two-dimensional visual still. I am both documenting and mediating my experience, at once connecting and separating myself from now, drawing closer and further from the material at hand.

Barn owls in tar seep.

Swirling black star. Sandy sediment. Feather and bone.

Natural asphalt. Manmade asphalt.

Before I was born, my father led photographic wildlife safaris through East Africa, camping for weeks across the plains and valleys of Kenya and Tanzania. I grew up surrounded by his photographs of lions, elephants, zebras, water buffalo, leopards, and cheetahs, framed throughout my childhood home. Each offered a portal: to the Serengeti, Ngorongoro Crater, Lake Manyara, Mount Kilimanjaro, Maasai Mara, Amboseli, Lake Naivasha. He had lived in South Africa as a scholarship exchange student during high school, Ghana during college, and after graduation on a Watson Fellowship spent a year with my mother camping in a Volkswagen bus up the southeast corridor of the continent from Cape Town, South Africa, to Nairobi, Kenya. Witnessing apartheid and colonialism during the 1960s, he started studying independence movements and economic sanctions on Rhodesia, to become Zimbabwe, alongside neighboring countries. He and my mother visited bush clinics and witnessed growing medical needs. His camera became a

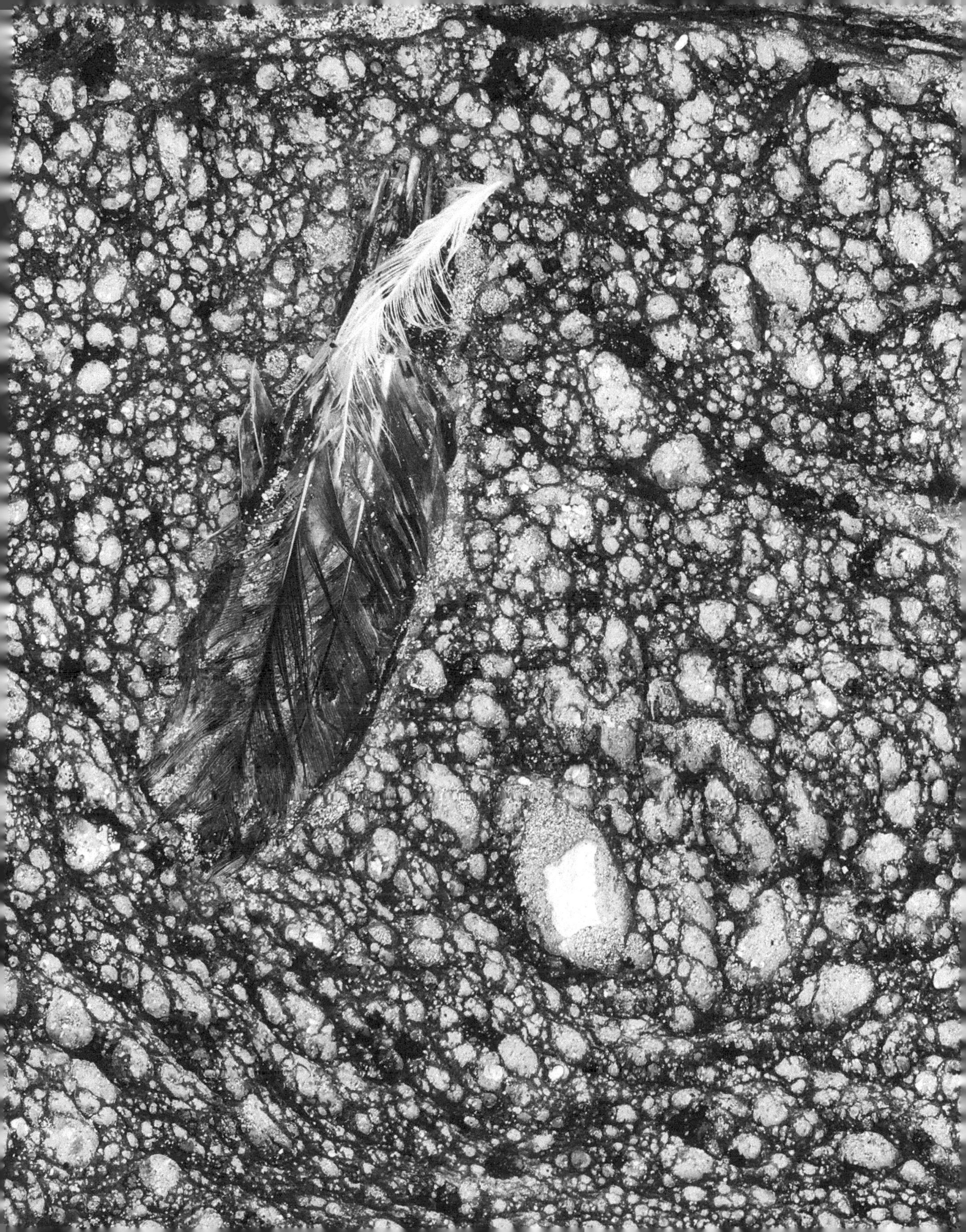

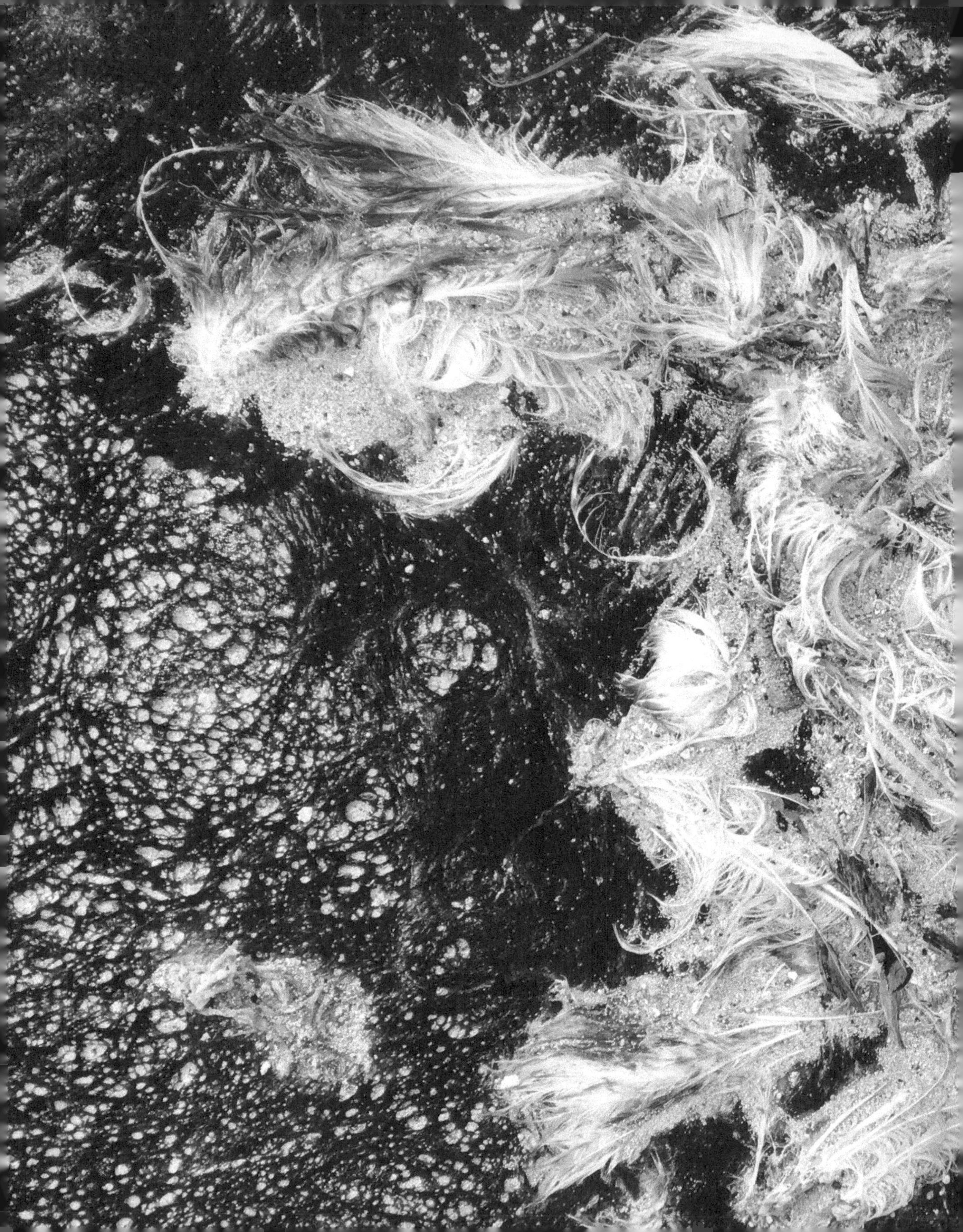

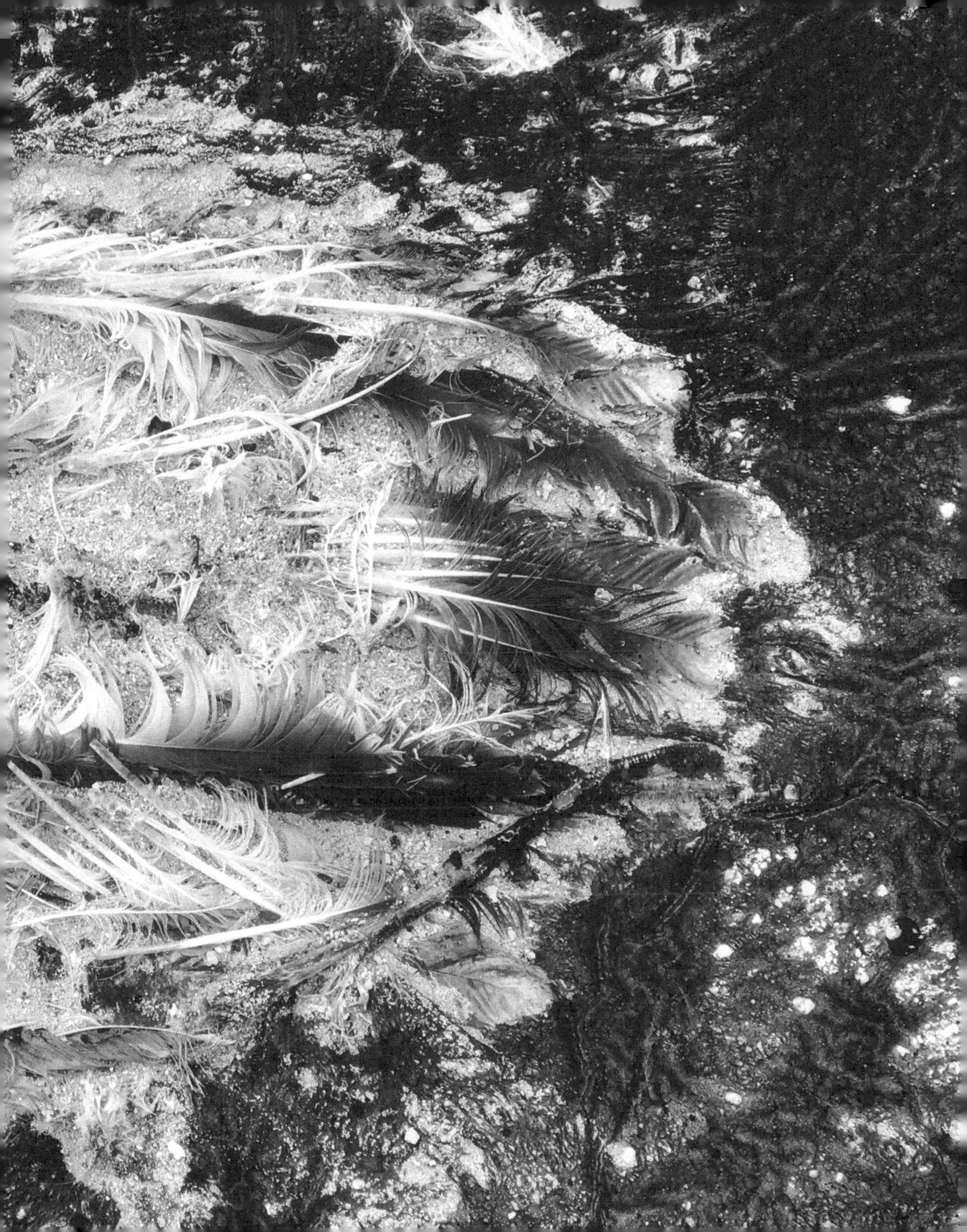

third eye. After returning to New York for graduate school in international affairs and night classes in pre-med, he spent summers leading photographic and camping wildlife safaris and sold some of his photographs. He once told me that if circumstances had been different, he might have taken a different path and lived somewhere in East or South Africa. But then came medical school back home in California, the births of me and my brother, my mother's tenured career as an epidemiologist, his aging parents: all of which rooted us deeper and deeper in San Francisco.

I take another photograph. Then, another:

Close-up fragment of bone

Wide-angle of the seeping star.

When I was born, my father stopped leading photographic wildlife safaris and instead led me on vicarious treks through slideshows in our living room. In my childhood home, I watched scrolling slides and imagined fog rolling off Table Mountain, inhaling the smoky mist of Victoria Falls, crossing borders to witness a new election, or camping by a watering hole. In our basement, he built a darkroom where the process of witnessing slowed down to stills. Photographic negatives turned positive through projections in his enlarger, washed in developing fluids and fixed in baths of light and shadow. We focused on exposures, saturations, gray scales, framing. On family treks to national and state parks, he was always the first to spot wildlife: black bears, blue herons, bison, egrets, elk. Despite his path to pediatrics in the Bay Area, he still seems most at home away from the fray, exploring and meeting locals, listening and asking questions, carrying a camera with an open eye on the horizon.

If I were to get stuck here at Rozel Point, I wouldn't know how to survive off the land and would merely begin a long trek back to Promontory Summit. There is no cell service. I am aware of the language that I have lost, that I am depending on words and images to articulate this disarticulation, rather than intuiting the multisensory world, attuned.

The wind chills my hands, as I try to grip my phone.

As I step forward to photograph the crevices of black tar edged with bone, the disarticulated feathers blur into white sprays like constellations.

Growing up in San Francisco, I never thought about why the skies above my urban neighborhood appeared mute. In the mountains, stars came alive and sat-

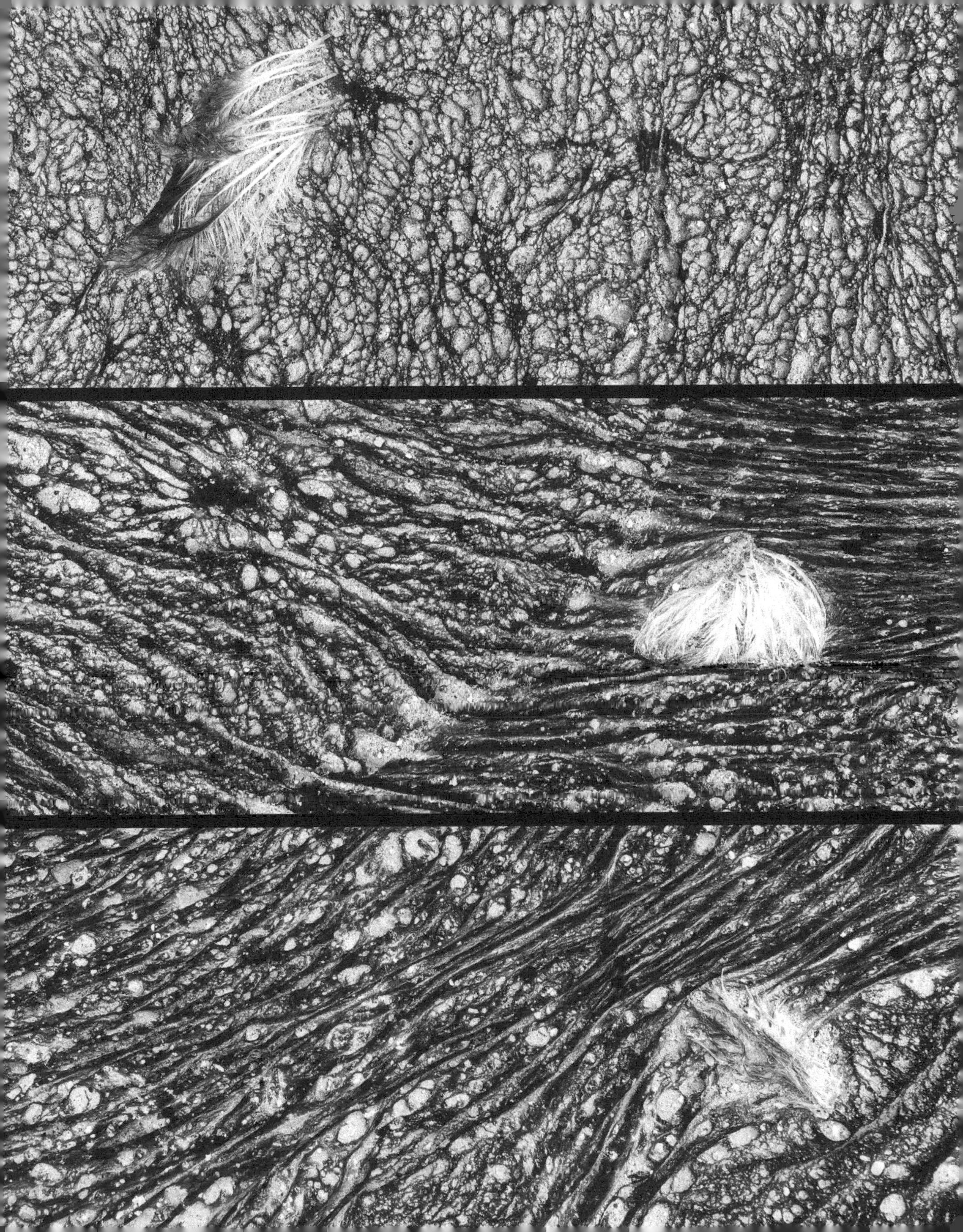

urated the sky, seeding my dreams to be an astronaut (until my fifth-grade classmates and I watched as, televised in real time, Christa McAuliffe's space shuttle exploded to a wisp). My awe grew in the Cascades and Sierras, where the night skies fluttered with constellations and showered meteors. When Los Angeles experienced the blackout of 1994, people were awed and terrified by the extraterrestrial swirl that emerged from the darkened sky—the Milky Way—usually hidden by light pollution. Our Milky Way is made up of approximately one hundred billion stars, so we cannot see the whole of our galaxy. We are perpetually inside its spiral.

By the owl remains, Greg lays down a measuring stick to photograph the bones. "Probably what you're getting is mice running out and getting stuck," he conjectures, "so it's like flypaper. The owls see them moving and come in to swoop them. If their feathers touch down, they get trapped." The mice may have been covered quickly because of their small size. "The owls are so big, it's taking longer for their burial."

Under the overcast sky, I imagine this place on a clear night. Far from artificial light, the Milky Way would swirl with stars. While it is estimated that many children born today won't get the chance to see the Milky Way, in Utah dark skies are making a comeback—the state now has the most designated parks for the night sky. In 2017 the University of Utah started the world's first academic consortium to study celestial protections. One of my graduate students spent last summer in Moab, inventorying over two thousand light fixtures in order to present recommendations to decrease light pollution to the city council. Studies show that light pollution not only obstructs the stars but throws off bird migrations and contributes to cancer rates. International Dark Sky Parks are planned to preserve night skies, reminding humans of our humble place on a living planet in a galaxy far larger than ourselves.

We keep circling the star-shaped tar seep, studying it from different perspectives.

Whenever I fly into the Salt Lake basin, my face presses the plane's window for a bird's eye view. Unlike the black tar seeps, the lake's salt evaporation ponds rivet through colors—turquoise, sea green, olive, sienna—in geometric fields like abstract paintings on the vast flats under the granite Wasatch Mountains. The scale is orienting and disorienting, as if those mineral ponds might lie under a microscope, floating in a Petri dish. The aesthetic is economic, signaling industries generated from the lake: salt evaporation for water softeners and road and plane

deicing; brine shrimp cyst harvesting for fish food; potassium sulfate for crop fertilizer; magnesium for auto parts, soda cans, cell phones, and pharmaceuticals. Greg later jokes that these are the lake's "liquid assets": the reason that the lake is increasingly endangered.

"Industries in the lake are important because money speaks," Jaimi says. "The flipside—without industry, could you support more birds?"

"In science, you never know the outcome," Greg adds, walking around the bird bones, "so you need to adapt to circumstances. Ask a question, pick up anecdotal data, and that gives you a starting point."

Like a massive ink blot on the planet, the tar seep seems to defy language, even as we grope to articulate its qualities. Greg uses the fossilized vocabulary of paleontology; Jaimi homes in on ornithology; others would describe the economic value of the oil. Without understanding the timescale of tar, I analogize it through

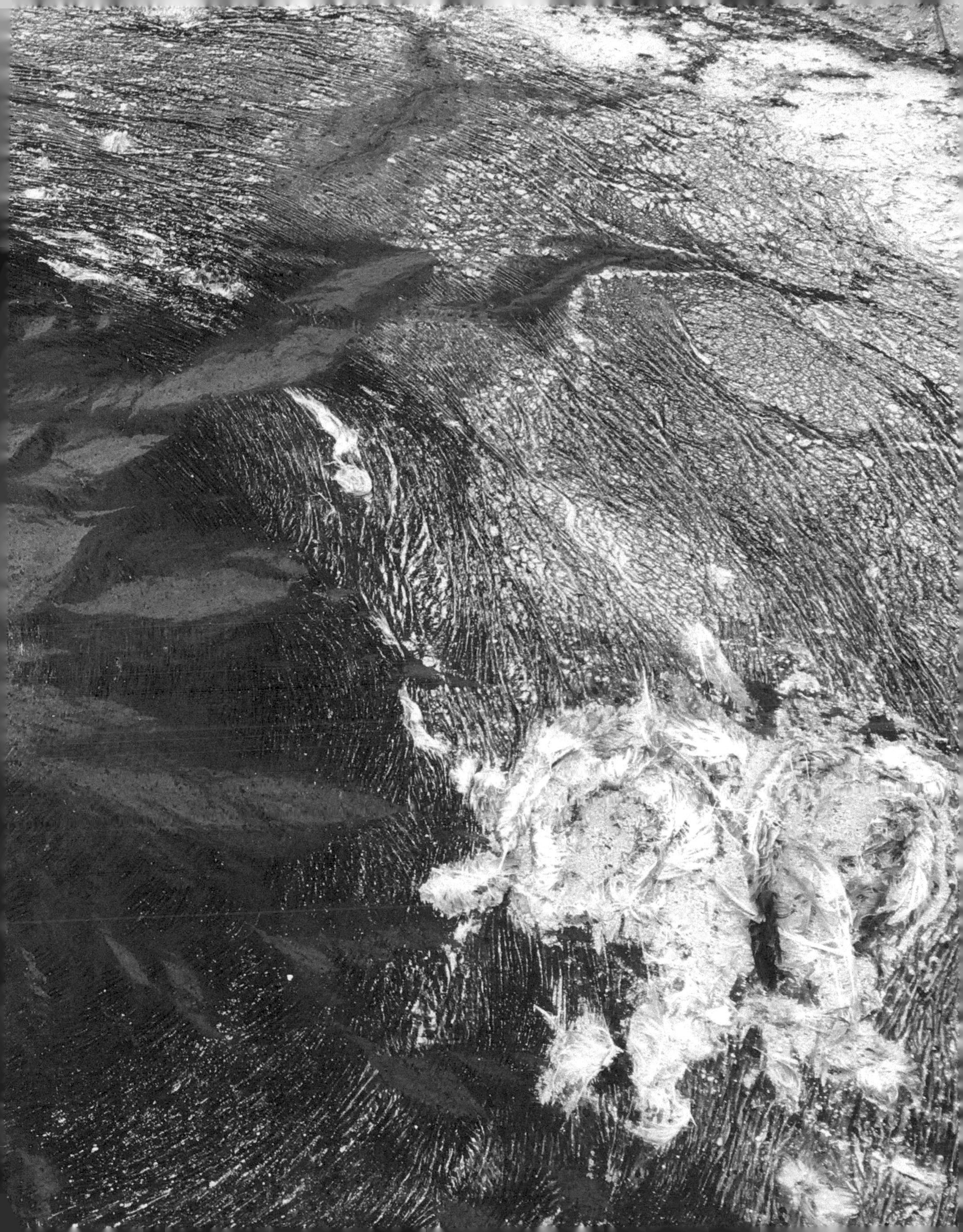

memory and associations. Even as I acquire terms, the seep evades classification. As tar sticks things together, it challenges a human tendency to classify: disarticulating anything that gets trapped.

Ultimately, the tar seep will swallow all languages that attempt to describe it. Bacteria break it down. Destruction and creation all rolled into one. As some bacteria even eat tar, they suggest hope for cleaning up future oil spills, and something more philosophical: about the power of microbial species, integral to our life cycles. Bacteria line our guts, maintain our body chemistry, and one day decompose us back to dirt—unless we get stuck in a tar seep.

On this frigid day, the tar is fully frozen. We could walk right across the seep. Yet we don't, instead circling its edges, as if the star might somehow stick us in place or collapse under our weight.

Later, I will return to the seeps in warmer weather and smell their melt, stepping in tar accidentally and leaving the trace of my footprints to be fossilized. I will see dozens of pelicans flying overhead. Later, I will question Land Art that inscribes the landscape. I will learn that charting microseismic waves of Great Salt Lake can reveal subsurface geology, akin to a CT scan of the Earth. Later, a geologist will lead me to Rainbow Bridge National Monument in southern Utah to listen to seismic vibrations of stones, where I will hear delegates from the National Park Service's Native American Consultation Committee for Rainbow Bridge, which includes Hopi, Kaibab Paiute, Diné (Navajo), San Juan Southern Paiute, Ute, and Zuni members. They will describe the reverberating sound as ancestral voices, leading me to stop taking stones as mementos, to leave them living in the places of our encounter, to try harder to listen.

But here and now, at Rozel Point, I hear only the wind. The tar seeps force our respect: to watch where we step. They invite *retreat* in the sense of *action:* to reconsider, to withdraw, to retract, to revoke, to consider nonintervention as a kind of interaction. Our presence is full of contradictions—driving a gas-guzzling van to do environmental studies as we try to "leave no trace."

On a gravelly recording from 1986 that I only recently found, my great-uncle Fritz's booming voice describes first hearing the word *ecology* in 1932, and learning more in the 1950s and '60s with the Earth Movement amid fears of PCBs, acid rain, and pollution. He unpacks the etymology of the word, from the Greek *oikos*,

literally "house" as home—meaning much more than architecture but rather a dynamic interrelation between where you were born and reared family, raised food, and died—a word that essentially asks: Where is your life? He describes moving away from fear into reverence for the totality of life, not separating the environment "out there."

I wonder, again, how Great Salt Lake remembers: as a body of water. It cycles through evaporation to snow and rain and runoff down the mountains back to the lake. Even though Fritz died almost a quarter century ago, for me he is one of those people who seems to live among the elements. I carry him with me like the gypsum in my pocket, an elder as a stone. "The [Ojibwe] word for stone, *asin*, is animate," writes Louise Erdrich. "Once I began to think of stones as animate, I started to wonder whether I was picking up a stone or it was putting itself in my hand."

I step back from the seep, and the gigantic black star shrinks to its actual width: about my height. In months these fragments may not be visible, disappearing into the earth. As I look at the owl bones, one fragment connects to a feather that reconnects to a wing into a body with a heart-shaped face. It is hard to know where one thing ends and another begins. I wonder about the lives of these four birds before their disarticulation, and what they will become. Beyond this lake, more birds migrate, near and farther than I can fathom. I wonder what compelled these owls to land on this tar seep and what led our motley crew to encounter them.

"Can we go?" Cora says, getting antsy.

Greg, Jaimi, and I look up from the tar seep. I realize that we have become like children ourselves, losing track of time. My fingers are numb.

After a final look at the seeps, we reorient ourselves to two distant rows of pylons that lead back toward the minivan. As we walk, Jaimi carries Cora on her back, talking with Greg about setting the camera traps with students in early summer. I feel as if we have looked through a telescope at a star, except instead of appearing in the sky, it has emerged underfoot.

Back at the minivan, Jaimi asks, "Have you climbed the bluff above *Spiral Jetty*?"

It's midday, and we are trying to return to Salt Lake City in time for me to teach my 4:30 p.m. graduate seminar. I doubt there's time to linger, but she recommends exploring the view from the bluff that helped inform Smithson's ideas of perspective. She still needs to check the road counter.

"I'm used to digging," Greg says, adding that he will help her.

"Look for the striped rock," she says, driving slowly to park. We all get out of the car.

"Be back by 1 p.m.," she tells me and pulls out a shovel.

I walk ahead on the rutted road. No one else is out here.

Looking up the bluff, I doubt there's time to climb and instead head down to *Spiral Jetty.* I start to follow the spiral, retracing my steps from October, but feel the tug of time. I double back toward the bluff and wind upward through sage-strewn rocks. The bluff's height is deceptive, so I rise quickly. Behind me, the lake's muted silver tints to rose and gold as I climb. I look ahead but glance back, watching *Spiral Jetty* grow and round out. The lake spreads to the seeming ends of the Earth. The sun breaks through patches of clouds. Later I will read Smithson's catalogue of the lake's reds: "wine-red," "tomato soup," "violet," "pink," "scarlet," "ruby," as "blood" from "the heart" with "veins and arteries." "Chemically speaking," he wrote, "our blood is analogous in composition to the primordial seas."

Smithson's written inventory of colors reminds me of another artist, Spencer Finch, who has indexed the lake's colors into a permanent exhibition at the Utah Museum of Fine Arts: an assemblage of 1,123 ready-made Pantone color chips, minimally presented as a band of thumb-sized swatches encircling a massive white gallery. The colors bear simple penciled labels—*lake, mountain, road, shrubs, grass*—reminding me how Smithson once published a list in repetition: "Mud, salt crystals, rocks, water." Finch's derived installation is subtle enough that on one visit to the gallery, I overhear a visitor who enters and says to her companion: "There's nothing in here." Later Jaimi tells me that Finch nicknamed the GSLI his "invisible hand," since the institute helped him to access the remotest parts of the lake. It's impossible to circle the whole, so they went on multiple trips—by land, by sea, by plane—as Finch tried to represent the landscape in color fields. Jaimi was glad that the GSLI could help. "When someone says they want to circumnavigate the lake," she admits, "I think they're going to die."

At the top of the bluff in ten minutes, I find a boulder and sit and stare out at the basin. This perch feels both in and outside time, as if all of the world has slowed to a stop. From this vantage, it is possible to peer offshore to Gunnison Island, where pelicans nest, and down shore to the tar seeps, where birds get fatally stuck in natural asphalt. They congeal in my imagination with manmade asphalt,

where I was hit by a car in a crosswalk, and with scars that accompany living. Like all living beings, the Earth also bears scars. Some visitors have critiqued *Spiral Jetty* as a scar. A few years after its construction, *Spiral Jetty* disappeared underwater (shortly before Smithson unexpectedly died) and didn't reemerge until three decades later when lake levels dropped. Smithson once described Rozel Point as a "time machine":

> The scale of the *Spiral Jetty* tends to fluctuate depending on where the viewer happens to be . . . A crack in the wall if viewed in terms of scale, not size, could be called the Grand Canyon. A room could be made to take on the immensity of the solar system. Scale depends on one's capacity to be conscious of the actualities of perception. When one refuses to release scale from size, one is left with an object or language that *appears* to be certain.

Perhaps that lack of certainty makes me feel certain today, strangely solid despite the slow-motion suction of the seeps. Each moment seems to resonate, like rocks that vibrate beyond human hearing. To sense this resonance requires listening beyond human registers, attuning to nerve and pulse, to bodies of water and of land, amid other bodies: animal and botanical and mineral. From this perch it feels possible to live across time and place, unfurling from the shore, seeping into the landscape, spreading across Great Salt Lake and to the West (to my home state of California and the Pacific Ocean) and to the East (over the Rockies, Great Plains, and Great Lakes with Lake Michigan, roughly the size of Utah's ancient Lake Bonneville), where my father-in-law spent the last decade and a half of his life trembling like an earthquake, bearing a disease named for a geologist who observed tremors in the earth. As a lifelong pianist, my father-in-law could calm some tremors by playing music stored deep in his muscle memory. To help him manage his Parkinson's, his neurologist repeated a mantra: "Expect the unexpected." Jaimi said something similar of Great Salt Lake: "The unexpected is part of the biggest thing for us."

Unfurling from the shore, *Spiral Jetty* evokes more spirals. The spiral is a shape of expansion. Once you see one, like a seep, you start to notice the shape everywhere: minispirals of stones that visitors have placed in the flats around *Spiral*

Jetty, chiseled spirals in rock art, a cyclone, swirling water down a drain. The shape is also unseen: infinitesimal as a microbial crystal or DNA helix, or cosmic as the galaxy of the Milky Way. Spiraling reflects space in time: a cartography of chronology. "One enters the *Spiral Jetty* backward in time," wrote the artist John Coplans, "bearing to the left, counterclockwise, and comes out forward in time."

For now, from the bluff on this wintry February day, the landscape is so expansive that it is possible to see far beyond *Spiral Jetty* to the horizon. I try to take a photo but can't find a flat line. The Earth curves, almost imperceptibly. Until you try to align the horizon, you miss the phenomenon. From this perch above the salty basin, the slight curvature provides perspective on what we're all standing on—less individual than collective, less dead than alive, always evolving.

Looking at the distant lake, I don't see a soul. A black truck drives along the road. It reaches the makeshift parking lot by *Spiral Jetty* and parks beside Jaimi's blue minivan. These two dots are the only signs of people in sight. I wonder why the truck has come. I look at my watch and realize that it is time to go.

. . . water around jetty • salt flat around jetty • rocks on jetty • rocks on jetty • mud on jetty • conifer shrub • sagebrush • salt flat • salt flat • salt flat • salt flat . . .

Descending the bluff takes little time. Within minutes, *Spiral Jetty* flattens and shrinks, coiling in on itself.

*. . . lake • lake • lake • lake • lake • lake—**At Spiral Jetty**—grass • grass • lake in distance • lake in distance • lake in distance • lake in distance • road • shrubs • grass • grass . . .*

By the time I reach the parking lot, the earthwork retreats into the lake.

*—**Road to Spiral Jetty South**—sagebrush • grass • shoreline • salt flat • mountain in distance S • lake • shrub • lake • lake • cloud shadow on distant mountain N • grass . . .*

The minivan sits in the parking lot. The striped rock is back in place over the road counter, with previous visitors tallied to the extent that they can be counted.

There is no garbage can or porta-potty, encouraging those who come to leave no trace of our presence.

I take one last look at the lake and head to the van.

When I open the door, Jaimi and Greg are laughing as they swap life stories. Jaimi trained in fisheries and wanted to be "the Dian Fossey or Jane Goodall of brine shrimp," while Greg credits his aspiration to a family vacation at Dinosaur National Monument. "Like a lot of kids, I caught the dinosaur bug," he says, chuckling. "I tell my friends who are dinosaur workers that I outgrew it and now work with mature stuff, like Ice Age animals."

Leaving *Spiral Jetty* the way we came, we follow the same road in reverse. A white truck drives toward and past us. It looks strangely out of place. The road leads along the lake, then turns inland. The lake's thin, shimmering line retreats in the rearview mirror.

As Jaimi steers the car through ranchland, I feel out of place yet strangely present. The same route always looks different from the other direction. Here blurs there. As we drive ahead on the road, it is clear that we cannot go backward in time, except in our minds, while experiences shore up inarticulately, non-linearly, in our bodies. Even if we were to turn around and return to Rozel Point, the lake would appear differently as the clock ticks forward, as Earth spins slowly ahead. The star-shaped seep of tar would still be frozen, but we would age by a few hours, seeing the seeps anew, as if two places at once, recognizing our presence there before. Like Stephen Hawking's broken cup putting itself back together in slow motion, we can't go linearly backward in time. "One can go readily from the cup on the table in the past to the broken cup on the floor in the future, but not the other way around," Hawking writes in *A Brief History of Time*. "The increase of disorder or entropy with time is one example of what is called an arrow of time, something that distinguishes the past from the future, giving a direction to time."

The lake is no longer in sight.

After my father-in-law died, I noted how his "struggling to hang on made us believe in the impossible: that little by little his breathing would improve and the process reverse, releasing the sickness in his system like a release of winds, cleansing the body until he would blink more alert, stop wheezing and gasping, starting to talk, first a whisper, then stronger and stronger, back to laughing and winking and talking, chuckling into a guffaw." Later, I will learn that molecules made by extremophiles in the lake and elsewhere are being studied to understand misshaped proteins, with hopes to treat diseases like Alzheimer's, Parkinson's, and certain cancers. As pesticides and pollutants shift human chemistry and activate genes that may have lain dormant, what tinges the lake pink may seep into our cells to color the future of life.

Jaimi points in the distance toward Promontory Point, describing a proposed industrial waste dump. A Class V permit was filed to allow controversial materials like coal ash (containing arsenic, lead, mercury, and other poisons) to be shipped by train and dumped near shore. Toxins that wouldn't be allowed in California could be dumped in Utah. If the liners leaked or winds blew, those toxins could contaminate the refuge. Since Great Salt Lake sits at the convergence of two flyways, migrating birds could ingest a toxin like mercury and die or take it elsewhere. The GSLI tries to educate the public: "It's not just a local problem."

"Coal ash, heavy metals, stuff with lots of bad juju—there's nowhere for it to go," she says. "What happens if that crap seeps into faults? We don't know what will happen or how aquifers and groundwater systems work in an 8.0 earthquake."

Although the landscape constantly shifts, Utah isn't prepared for the overdue tectonic threat. The surrounding terrain is largely shaped by earthquakes. Google Earth shows a bulge in the railroad causeway across the lake, where rocks have squished to the sides. The causeway divides the north and south arms of the lake, whose halves appear as different colors since the north is strangled from a freshwater source. The salinity level of the north arm can be as high as 27 percent, while the south arm averages half that. Oceans like the Pacific average closer to 3.5 percent.

"I'm afraid that they'll give up the north arm for dead," Jaimi sighs. "Kill the north arm to save the south arm. People have been trying to kill Great Salt Lake for years."

It's hard to imagine that human-placed stones like the causeway can change the lake's biochemistry. I ask what locals thought of Smithson moving boulders into the shape of a spiral.

"It was basically one person: the guy who drove the backhoe," Jaimi laughs. "He didn't really understand what Smithson was doing until years later when his daughter showed him a cover of *TIME* magazine with *Spiral Jetty* on the cover.

Tremonton
Provider
Honeyville
Hyrum
Promontory
89
15
Bear River
Migratory
Bird Refuge
Spiral Jetty
Earthen sculpture in
the Great Salt Lake
Great Salt Lake
Ogden
Roy
89
84
Clearfield
Layton
15
Antelope
Island
15
Bountiful
80
Salt Lake City
gonite
80
80
West
Valley City
Erda
Grantsville
West Jordan
Google
Sandy

Remember, it was covered up for years after Smithson made it." She explains that the meandering zone of lakebed on which *Spiral Jetty* rests is leased by the Dia Art Foundation from the State of Utah Division of Forestry, Fire and State Lands.

In 1960, a decade before Smithson made his earthwork, a proposal was mounted to make Great Salt Lake National Park. The bill didn't pass, since the lake had been subdivided into public and private, corporate and government hands. Over a century earlier, Stansbury had circumnavigated the lake using smoke fires. Remains of one of his triangulation stations can still be found on Gunnison Island. Lack of fresh water prevented homesteaders from settling Gunnison (except for a short time by Alfred Lambourne who brought a piano, planted grapevines, and wrote a book about his misadventure called *Our Inland Sea* in 1909), but in recent years, as the lake level has dropped, land bridges have started to emerge. Reaching Gunnison Island, coyotes and ATVs can roam and scare off pelicans from their breeding grounds. The GSLI watches these patterns by traveling annually to the island with the Utah Division of Wildlife Resources, tagging pelicans and studying how different parts of the ecosystem connect: through birds and brine shrimp, halophiles and gypsum, salt minerals and fault lines. It's not one environment but many microenvironments, interconnecting across the north and south arms, through water columns and currents, around the lakeshores and the water cycle of the Basin and Range.

When we pull into the parking lot at the Golden Spike, there is only one other car. We use the bathroom without going into the visitor center. Back in the van, I ask Greg what it is like to work for the Bureau of Land Management at this moment in history, given threats to national monuments and public lands, and how this compares to his lifelong work.

"They've streamlined the process without talking to the EPA (Environmental Protection Agency)," he says. Loosened protocols have made it easier for companies to drill and extract natural resources. As the bureau's regional paleontologist for the West, Greg is professionally based in Utah but travels between field offices to evaluate fossils and teach paleontology, covering a number of states, including his homebase of Colorado. He was formerly a curator of natural history collections for the National Park Service and has consulted on tar pits and tar seeps around the world. Later when I travel farther through Utah, I will learn indirectly of his admired outreach and educational commitments. "Years ago," he laughs now, "I realized something working for the federal government: If everybody's mad at you, you're doing something right because you're not showing favoritism."

The two-lane road bypasses low hills pocked with limestone caves. The caves formed when the lake receded. Limestone precipitated out and left hollows. Paleo-Indians inhabited the caves, leaving thousands of objects that archaeologists study to piece together previous communities. Caves also show continuous owl habitation through pellets, expelled with the fur and bones of their prey. In other parts of Utah, rock art depicts owls.

"Lake Bonneville has gone through many cycles," Greg explains. "Great Salt Lake is the latest incarnation in this lake history. If you drill down to the bottom, you get these evaporite deposits—salt, gypsum, stuff like that—when water concentrated down, and lake levels dropped." Each phase gets a different name to convey the dynamic of time. "This is not a static system; it's very dynamic. The question is: How old is Great Salt Lake? Drilling down, there are suggestions that versions of the lake existed millions and millions of years ago and, in some cases, totally dried up."

I look out at the watery marshland.

"There are suggestions in the older rocks that the lake was here," Greg continues. "Then it totally went away. Then conditions came back, where water started accumulating again."

I imagine the water receding, all the birds flying away. Or trying to stay and dying. A dry and birdless lake.

"Look at Great Salt Lake today," Greg continues. "Look at how it's going down with these dry conditions. You're not getting the flow. We know that during dry periods, the lake goes. It's theoretically possible that you could have Great Salt Lake dry up during our lifetimes or within the next generation."

Time unfurls in my mind in fast-forward: the seep of a black star melts; pelicans fly away; the lake no longer appears as a still photograph, rather a film blurred by a dust storm that thickens to dense darkness.

"It's happened in the past. Look at Lake Manix, Lake Searles, Lake Lahontan. Look how many other Great Basin lakes are just dry playas now. Given what's going on, Great Salt Lake could become the Great Salt Lake Playa if we keep monkeying with the water supply."

As we drive by the Bear River Migratory Bird Refuge, it's hard to imagine the loss of birds and depleted flyways. It is also hard to imagine the lake evaporating entirely—the age-old bathtub draining to dust to leave another dry lakebed in the arid West. But perhaps more startling: it is also easy to imagine it. The southern part of Utah was once a massive desert, much larger than the Sahara. When water was diverted from Owens Lake in California a century ago to fill the Los Angeles Aqueduct, the lake shriveled and has now become the largest source of dust pollution in the United States. Great Salt Lake might follow a similar course. The Bear, Jordan, and Weber Rivers that feed the lake with freshwater are increasingly diverted for agriculture and polluted. A lower lake level leaves less water to evaporate into the water cycle and return as rain or snow; less snow in the mountains yields less runoff; dust storms quicken snowmelt; and the cycle continues.

As the lake level drops, other matters rise to the surface. Jaimi recounts how when she first worked on the lake, counting brine shrimp, an unofficial duty involved delivering wine to the harvesters, or "brine shrimp cowboys." Unbeknownst to her, they put her name and number in the wine bottles, with instructions to call her as part of a scientific study. One day a police officer called and asked if she was in trouble, since a bottle had been found near partial remains of a human body. Only later did she learn that the remains may have belonged to a four-hundred-year-old Native American woman. She doesn't know what happened to the remains or who else was contacted.

Almost imperceptibly, the air smells like rotten eggs.

"It smells like science," Jaimi says, repeating what she tells school groups.

As we speed down the highway, it is clear that Great Salt Lake holds layers and layers of stories, told and untold. Jaimi shares her shorthand for the lake: *seagulls, shorelines, salt,* and *stink*. Others would describe the inland sea in cultural, artistic, or economic terms. Through a scientific approach, Jaimi works with teachers to develop locally based curricula, currently focusing on colors, while encouraging collaborations to better understand the lake's shifting story. Students and teachers come to *Spiral Jetty* to see, feel, smell, and touch the landscape to appreciate its environmental complexity. The more questions are asked, the more gets uncovered. The difficulty of the drive contributes to Rozel Point's lasting impression, especially if your car gets stuck in a rut from a flat tire, leaving you stranded for hours.

As we drive farther from the lake, the snowcapped peaks of the Wasatch grow. It becomes clearer how Rozel Point with its tar seeps connects to everything around: Great Salt Lake, the Great Basin, North America, and beyond. Rozel Point is a microcosm of interests—scientific, artistic, cultural, economic, political, more—at the intersection of public and private lands. At a glance it may be easy to dismiss the lake as lifeless or dead, stinky or ugly, until we listen: to presences and absences of cultural histories, to those who care about its halophile-tinged, saline waters, and to many other forms of the lake's life. Extractions at Rozel Point interconnect with other parts of the state, where loosened protections of national monuments, like Bears Ears and Grand Staircase–Escalante, promote drilling and fracking and threaten the watershed of the Colorado Plateau and Native communities who inhabit ancestral grounds. As environmental protections are rolled back, the Earth becomes more vulnerable to injury—as large swathes of California and other Western states have burned after being riddled by bark beetles and drought, as monumental hurricanes have flooded the Southeast and decimated island nations like Puerto Rico. The surreal reports keep coming around the globe: from melting permafrost across Alaska, to a real Northwest Passage emerging as Arctic ice melts, to South Africa's announcement that it will run out of fresh water in a few weeks, to a mammoth crack in Antarctic ice. Geologists have found that manmade reservoirs can seep into cracks, lubricate earthquake fault lines, and jumpstart the forces of the planet. It becomes clearer that attending to only one issue or to a single species alone jeopardizes everyone. Everywhere is here under a shared atmosphere. Our individual actions press into lives beyond our own—not only human and animal bodies, but also bodies of land and bodies of water—resonating in ways that, if listened to, might helpfully decenter our individual selves from the center of any story.

As Jaimi steers the minivan onto the interstate, Greg explains that "paleontology is a historical science" around two centuries old. Beginning his career before the Internet, Greg has spent decades collecting rare books on vertebrate paleontology, comparing illustrations to actual bones in museums: by Georges Cuvier, the so-called father of vertebrate paleontology; by Charles Darwin when he sailed on H.M.S. *Beagle;* by Sir Richard Owen from South America. "I may disagree with his interpretations, but his anatomical descriptions and information are still relevant," Greg says. "I like it because it gives context."

As Greg talks about his work curating natural history, I feel like I am in two places at once: this car driving back to Salt Lake City today, and at the Smithsonian last semester, visiting bird specimens to piece together a lost museum from the university where I seasonally teach in Washington, D.C. For more than a century, birds were collected, stuffed, tagged, displayed, donated or discarded, moving from field expeditions to prominent glass cases, drawers, basements or rubbish bins on campus. Those birds at the Smithsonian that survive posthumously, even post-extinction, are studied through their DNA. Each bird came to reflect the humans who searched for them: time capsules of knowledge in different eras, often classified into teaching and research, suggesting how we rationalize more than empathize with this planet that will go on without us if we don't proceed with care.

Our return to the city is both long and quick. Timelessness gets stuck back in time. The spires of the Salt Lake Temple rise. Under the granite fins of the Wasatch, the capitol dome flares in the late afternoon sun. Off in the distance, the Bonneville Shoreline Trail cuts across the face of the mountains. Time clicks back into place, getting me to class on time and into the motions of the week, as more weeks unfurl, as my classes blur into plane trips out of the basin to Indiana, Massachusetts, Florida, D.C., California. The weeks roll forward, and my students and I meet with experts on stones, gaining more vocabulary to talk about the same thing, which always changes. Whenever I leave the airport to fly over the lake, I look for Rozel Point, for *Spiral Jetty* and the tar seeps, but the planes' flight paths usually rim the south arm of the lake, flying at a distance from the site that I squint to see.

lake
lake
lake
lake
lake
lake

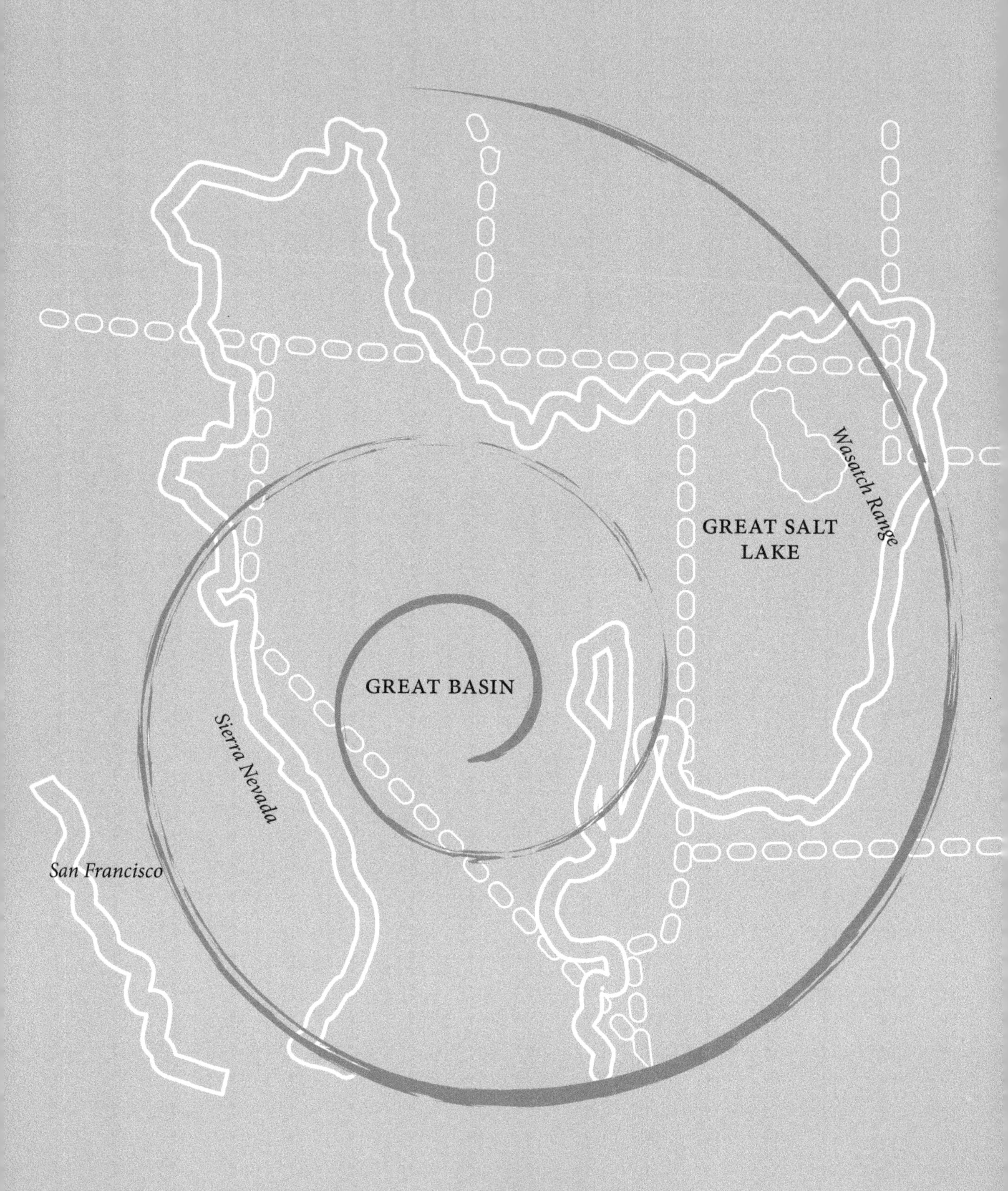
GREAT SALT LAKE
Wasatch Range
GREAT BASIN
Sierra Nevada
San Francisco

AMERICAN WEST

II.

To take a photograph is to participate in
another person's (or thing's) mortality, vulnerability, mutability . . .
All photographs testify to time's relentless melt.
—Susan Sontag, *On Photography*

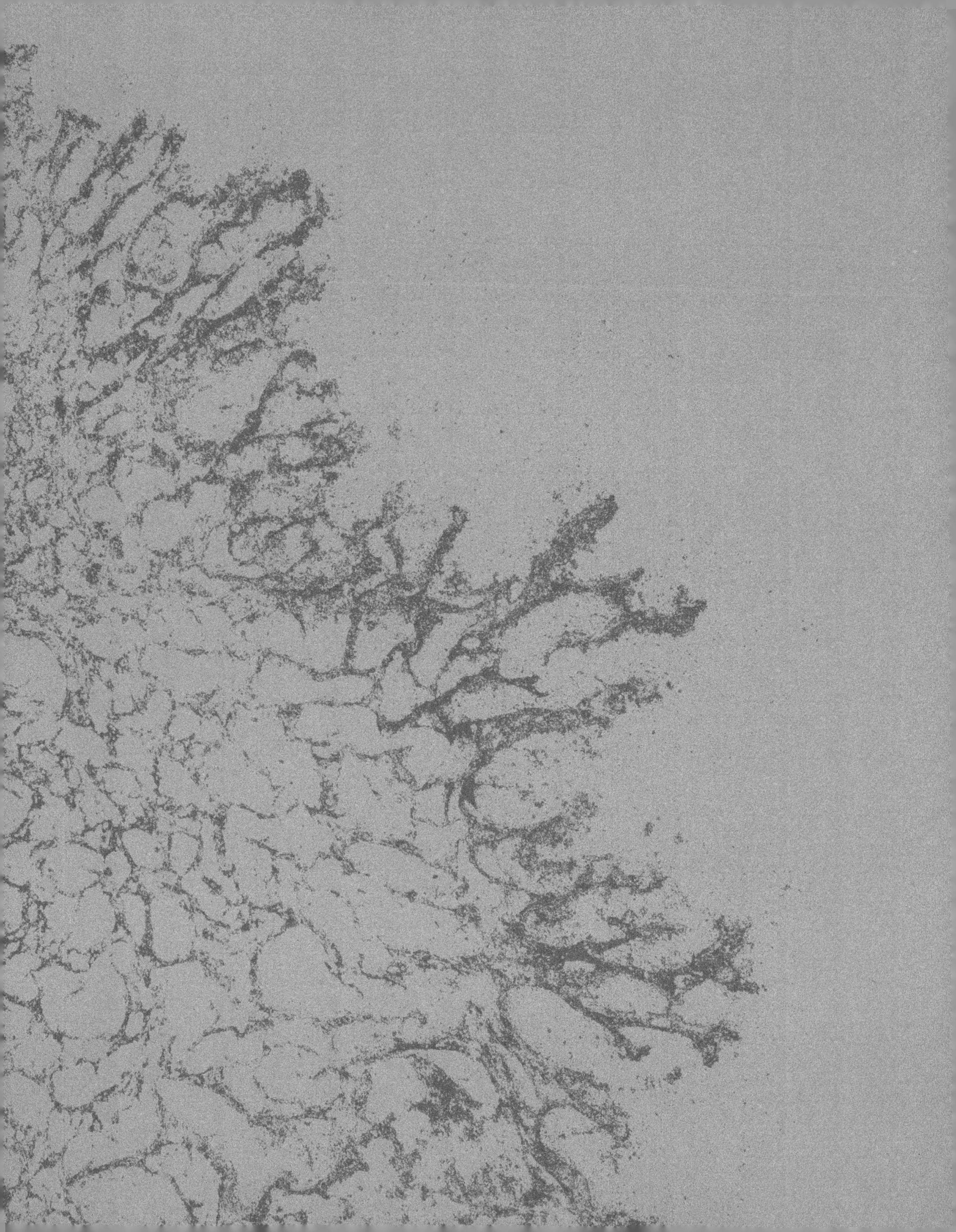

Stuck

With its star-shaped tar seep, Rozel Point is one of those places that sticks in my thoughts as weeks roll forward on the edge of Great Salt Lake.

On campus in the South Physics Observatory, the building's telescope focuses for me to see the Moon's *lunar maria:* large basaltic plains named ethereally after being mistaken for dead seas: *Sea of Tranquility, Lake of Forgetfulness, Ocean of Storms.*

When I travel east with family who are visiting, dark skies spread above the Uintah and Ouray Reservation and Dinosaur National Monument, with its tilted riverbed of prehistoric bones. Uplifted with the Rocky Mountains and later excavated, the parched land reveals time at a slant. Nearby oil fields light up like a metropolis, as each drill gets its own nightlight. Billboards beg to stop the grab of public lands for extractions of natural resources.

Throughout Salt Lake City, signs in yards read: "Utah stands with Bears Ears" and "Save Grand Staircase–Escalante." Regional debates rage over redlined districts, as the state's voting record suggests less care for Native American and environmental rights.

When I travel west, Interstate 80 rims the southern edge of the lake. My students and I follow that route for a field trip to Nancy Holt's *Sun Tunnels.* As both an artist and Smithson's wife, Holt shepherded *Spiral Jetty* and his career after his plane went down in Texas as he was surveying the site for *Amarillo Ramp,* killing him at age thirty-five. Since our trip to the *Sun Tunnels* takes a full day, time is crunched. We take a stretch break at the Bonneville Salt Flats but, short on time, need to bypass the Center for Land Use Interpretation and Historic Wendover Airfield (with its museum that includes the *Enola Gay* Hangar and an origami crane made by a short-lived survivor of Hiroshima) with hopes to return in the future.

Back in Salt Lake City, gulls remind me of my native San Francisco. The birds' cries reputedly led them to be named from an old Breton verb, *goelaff,* "to weep." At Tracy Aviary, the largest and oldest free-standing aviary in the United States, I find barn owls with their signature heart-shaped faces, golden and brown feathers. They hunt at night, feed on rodents, and triangulate sound to locate prey. Barn owls reputedly have the best hearing of any animal. Where farmland is being developed for subdivisions, they are experiencing habitat loss. I recall bird bones at the tar seeps, weathering and swirling into the fossil record.

Another day at Great Salt Lake, on Antelope Island, a friend shows me more birds that inhabit the basin: meadowlarks, chukars, rock wrens, long-billed curlews. It helps to have more words to distinguish flight paths and migrations.

My flights over mountains frame a landscape stamped with dots and curlicues as acres upon acres of fracking. Through canyons of the West, the Green and Colorado Rivers carve paths across the Colorado Plateau to water most cities of the Southwest, draining toward the Gulf of California and the Pacific Ocean. Like concerns for Great Salt Lake, there is worry that their waters will run lower, with fewer nutrients and more contaminants. Other concerns spread across the country.

Wherever I go, time curls, as *Spiral Jetty* and the tar seeps reorient my sense of place. Great Salt Lake stretches my imagination beyond Utah. As I fly in and out of the basin and across state lines, the lake pulls—as a beacon, focal point, harbinger, warning, the eye of a sleeping giant on the Earth. Associations layer. A geologist teaches me about arch song, or vibrations in rock formations, and about the lake's microseisms, or tiny waves that hum beyond human hearing. Listening to recordings of red rock arches stills my pulse and breath, as if awakening a memory beyond my lifetime. The sound is little understood by humans but resounds as if Earth has a heartbeat. By averaging seismic waves over time, a signal starts to emerge.

When my husband and I travel south, red rock canyons and arches seem to vibrate in the baking heat. Zion, Bryce, Grand Staircase–Escalante, Capital Reef, the San Rafael Swell, Goblin Valley, Little Wild Horse Canyon, Dead Horse Point, Castle Valley—the rock is full of iron, the color of blood. In Zion's Echo Canyon, two ravens wing through tight rock walls, amplified so loudly that the sound flutters my skin. Beat by beat, my pulse slows. The Swell washes out like lunar seas. In gradients of gray and rust, the dry landscapes appear extraterrestrial, even hosting a research simulator for Mars.

Later, after I have moved back East, a sizable earthquake will shake Utah. The earthquake will strike with repeated aftershocks as a pandemic sweeps the globe. Casualties and hardships escalate everywhere. The climate crisis deepens to a seeming point of no return. Narrative patterns keep spiraling: into collective loss, grief, outbreak, catastrophe, apocalypse, bleak future predictions. Yet simultaneously, voices amplify among interrupted industries, suspended commutes, and peaceful protests against racist violence. Birds seem to sing louder as streets empty of cars, as people notice what was always there, as skies temporarily clear of pollution, as resistance grows in slowed time: against inequities entangled with overused resources, slow violence and climate injustice, where imbalances are not fated, as shapes of stories can shift with many possible endings: not yet finished.

How do we repair injurious pasts in the present to renew potential futures? How do we not fall back on worn narrative patterns? How do we reimagine a story while it is being told?

Rewind: In early photography of the American West, landscapes often were portrayed as pristine wilderness, as if people had never been there. An ethos of "virgin" territory framed colonial settlement to perpetuate the destruction and dispossession of Indigenous peoples and lands, to shift White supremacist attentions away from legacies of enslavement of African Americans in the agrarian South. As new industries extracted more natural resources, humans increasingly separated from natural environments that sustained them, throwing off balances of land use and replenishment. Photographers who traveled with expeditions at times positioned cameras deliberately to avoid capturing images of living Native civilizations and cultivated lands, reconstructing the Western landscape as barren and dramatic in its desolation. Acts of surveying, naming, mapping, and photographing overwrote and reframed previous knowledges, displacing and replacing senses of place. Mili-

tary forts defended territories, then political states. Historical accounts reinforced patterns: of progress, of frontier, of settlement, of homesteading, of pioneering, of national expansion, celebrated as conquests that silenced other stories. Disregarded regions were carved into reservations for Native Americans, internment camps for Japanese American citizens, and testing grounds for nuclear bombs whose fallout led to cancers in "Downwinders." As mass travel increased, time sped ahead, delineating before and after, cause and effect, there and here. Injustices perpetuated. Place and time became mythic as disembodied data: as if place could be reduced to a GPS coordinate, as if time could be adequately tallied by a clock or summed up by a book, like this, held in your hands.

esources a
t of reside
sis in the d
ds. Songbir
aces as M
ater to drin
abitat alt

quired a le

t and migra

esert" and i

ls, e or

xico, South

k. Here t

ernatives fo

Fast-forward: in Salt Lake City I travel by foot as far as I can. Walking takes time as my main mode of transport: downtown, through neighborhoods, along creeks, up trails into hills, where vistas reveal a grand expanse of Great Salt Lake. Grateful for car rides farther afield, I want to visit many places in the region but do not get the chance, given personal limits. Imaginary and recollected travels arise through books, films, archives, artworks, and other intermediaries. I read about a temporary installation, *SALTWORKS*: made in 2014, in partnership with the Great Salt Lake Institute, when the Japanese artist Motoi Yamamoto made intricate labyrinths of salt, ephemeral as Tibetan mandalas or Diné sandpaintings, completed when that salt joined the lake at *Spiral Jetty.* I see a documentary film on *Repel-*

lent Fence/Valla Repelente—an aerial line of massive yellow balloons with scare eyes crossing the U.S.–Mexico border—made in 2015 by the art collective Postcommodity (Cristóbal Martínez, Kade L. Twist, Raven Chacon) as a healing "suture," critiquing Land Art from a "Western scientific worldview" that "constructed borders behind and in front of it," and countering the "colonial vanishing trick" that has erased Native perspectives and deprivileged integrative knowledge systems through dispossessed lands. Near the Salt Flats near Wendover, the Center for Land Use Interpretation hosts an interpretive station to question how the "man-made landscape is a cultural inscription, that can be read to better understand who we are, and what we are doing." Since 2017 in "Partially Missing Building," the art-

. . . grass • distant shoreline • grass • mountain in distance W • water in swampy area • lake in distance • grass • grass • shrub . . .

ist William Lamson installed the ongoing, time-based *Mineralogy,* while a nearby exhibit displays an origami crane made by Sadako Sasaki, a short-lived survivor of Hiroshima. Since the early 2000s, the ever-growing collective of Land Arts of the American West has camped on educational-artistic-cultural pilgrimages to Land Art and other earthworks, Native American archaeological sites, open-pit mines, bomb testing grounds, radio observatories, other human and geomorphic constructs and communities, growing dialogues en route. More spirals outward.

The more I learn about spirals and seeps, the more I question places where I get stuck. "How do we begin to deal with the 'problem' of Land art?" wrote art critic Emily Eliza Scott. "How do we become sensitized to the often unconscious, ongoing forms of colonization reflected in our own scholarship and teaching, including the frequent tendency to impose, encircle , simplify, lay claim, instrumentalize, and so on, as a means to consolidate power?"

I wonder about degrees of inscription: if inscribing words, including for this book, may not be so different from paving a road or drilling an oil well, by degrees: drawing borders or extracting materials to make something else legible. Electricity powers my computer; mass-produced paper and ink feed my printer. Felled trees lie in these pages that hold my text. Disciplinary and colonial languages, from which I have learned, separate me from the natural world while attempting to integrate, at times reinforcing hierarchies that claim to aim to dismantle. Inscription can be delicate as a mark on a page, as footsteps trailing in dirt, as bodies that cross paths or collide.

When I think back to Spencer Finch's installation at the Utah Museum of Fine Arts that virtually circles the lake, his color swatches outline absences more than presences, with few birds and no humans. *Sean's car* is the only reference to a person, alongside scarce animals: *deer, bird(wing), bird(coot?), fox, raven(crow?).* When I initially read *Sean's car,* it startled me by drawing my attention to the lack of people, and I misread it as *Sean's ear.* (*Who is Sean?* I thought. *Why did Finch represent just his ear?*) Maybe I wasn't alone in misreading. Perhaps someone else saw the handwriting inscribed as Sean's askance *eye.* Variations grow in the margins like illuminated vegetation, coiling around sources of where any story lives.

When I circumambulate *Great Salt Lake and Vicinity,* I sense the gallery's borders, walls, and doors—but outside, seeking the actual lake, I never discern the water's bounds. The lake slips around islands and the base of mountains, in and out of view, glinting in the distance. Like the unfurling *Spiral Jetty,* the lake decenters. As I look, again and again from different angles, time and space grow around incomplete aspects, here and now, within my body: not in isolation but in relation. Inhaled through particles of air, invisibly, the lake enters and decenters my very breath.

A breeze cools my skin. Flash of sun. A bird's wing (today: a magpie) shifts my attention to the mountains, to the lake, tracking my attention: grounding me in the present.

To represent Great Salt Lake is like trying to bottle up weather or catch the wind: the lake eludes the ability to be grasped, as it rides the curve of the Earth. Around the melting and freezing tar seeps, around *Spiral Jetty,* the lake moves in seeming stillness. Microseisms emanate as faint tremors from the slowly moving Earth. Storms whip up waves and seiches. Even on a seemingly still and sunny day, there is movement: buoyant in dense saltwater, evaporating into the atmosphere. Vapor condenses and precipitates, as storms surge across mountains into the basin. Weather unsettles the sky. Snow melts downriver to mix freshwater into the saline lake. Birds flock to marshes. Animals track around habitable edges and islands. Microbial life swims, finless, in colorful saline saturation, immune to the burning sun. Tectonic faults unleash larger shifts, shaking up manmade and natural settlements.

As a decomposing platform, *Spiral Jetty* offers a focal point to this climatic movement. Its path of salt-encrusted, black basalt spirals from the shore, outward to the seeming sea and curving horizon, then in reverse back to shore. For three decades after its creation, the artwork lay unseen, underwater. When it arose from the receding lake in 2002, it grew from a mythic absence, a phantom of presence. It echoes a mythic whirlpool while being a bellwether of drought. Seemingly larger than life, in reality the artwork is quite small. Photographs exaggerate its size and scale, centering what otherwise decentralizes. Its spiral grows at a distance. In aerial photography, *Spiral Jetty* becomes the focal point. Standing on *Spiral Jetty,* the view turns outward: to the lake, to the mountains, to clouds and sky. A walk on the artwork unfocuses its shape; its foothold of rocks invites a viewer to cycle through the expansive landscape.

As with other Land Art, photographs cannot replace the work itself. Attempts to represent or border it—across photographs, film, writing, archives—reveal the

. . . grass • grass • hillside • hillside • grass • grass • mountain E • mountain E • lake in distance • grass • salt flat • salt flat • salt flat . . .

limitations of whatever medium, expose the view of a creator, lose what lies at the edge or outside the frame. *Spiral Jetty* is never singular: crystallizing, decomposing, fossilizing. The artwork enlivens over time. Both prehistoric and futuristic in reach, Smithson constructed spiraling contexts around the earthwork: through a film, writing, photographs, maps, drawings, diagrams, documentation, archives, and an unrealized museum. Located through dislocation, his sites and non-sites constellated over years with other Land Art inscribing the American West: from Nancy Holt's *Sun Tunnels* (1973–76), to Walter De Maria's *Lightning Field* (1977), to Michael Heizer's *City* (begun in 1972 and still in progress), to Postcommodity's *Repellent Fence/Valla Repelente* (2015) that ephemerally, aerially, non-invasively stitched together the U.S.–Mexico border.

Moving outside galleries into vast open spaces, often in the American West, Land Artists integrated architecture, engineering, and natural canvases inspired on a scale of ancient Egyptian pyramids, England's Stonehenge, Native America's Great Serpent Mound, or Peru's Nazca Lines. The historical backdrop included bombscapes of the Vietnam and Cold Wars, the Space Race and science fiction, civil rights and ecological protests. *Spiral Jetty* was constructed within a year of the massive oil spill off the coast of Santa Barbara, California (near the Carpinteria Oil Seeps and Tar Pits), whose devastation led to the passage of the National Environmental Policy Act. The first lunar landing yielded photographs of Earth: small as a swirled marble. Perspectives shifted. Bulldozers were likened to paintbrushes. Earthquakes become fellow sculptors. Time and weather became collaborators, as did widening frames of reference: around archaeological and modernized human settlements, large-scale excavations and extractions, military-industrial installations, interplanetary simulators and radio observatories, tuning to listen to the minispiral of gas at the center of our spiraling galaxy.

In his film about *Spiral Jetty,* as Smithson traverses Rozel Point, the former tower and equipment around the oil jetty appear in the background. *Utah Sequences,* a long-lost short film by Nancy Holt from that time, also shows drilling equipment alongside tar seeps and dead pelicans. Oil drilling at Rozel Point lasted the better part of the twentieth century and dwindled in the 1980s around the time that lake levels rose. In his writing on *Spiral Jetty,* Smithson identified the tar seeps as central to his selected site. Asphalt was the material for his proposed 1966 earthwork for Philadelphia, *Tar Pool and Gravel Pit,* and his other proposals included *Asphalt Spiral, Series of Eleven Asphalt Pavements,* and *Asphalt on Eroded Cliff.* A quarry near Rome, Italy, became the site of his 1969 sizable melt, *Asphalt Rundown.* In an essay on "A Sedimentation of the Mind," Smithson wrote of his attraction to the material of "tar" that "makes one conscious of the primal ooze . . . This carbonaceous sediment brings to mind a tertiary world of petroleum, asphalts, ozocerite, and bituminous agglomerations."

In 1999, after Nancy Holt donated *Spiral Jetty* to the New York–based Dia Art Foundation, the lake receded, and the artwork reemerged: raising questions about its care. By 2005, the road was improved and most equipment around the tar seeps was cleared away: pump jacks, tanks, pipes, boilers, tubing, gathering lines, dilapidated cabins, even a rusting military amphibious vehicle. Yet the site remained remote, fairly inaccessible. A threat of oil drilling in 2008 revived efforts by Nancy Holt to protect *Spiral Jetty* and to steward remote earthworks. Dia joined efforts with local partners who could more regularly access the site: the Utah Museum of Fine Arts, Great Salt Lake Institute, and Utah Division of Forestry, Fire and State Lands.

The removal of ruins around *Spiral Jetty* arguably smudged the "missing link" around which Smithson speculated. Art historians wonder how Smithson may have felt about the mining equipment's removal. Like the industrial ruins of his homescape of New Jersey, or irrigation equipment in Texas just outside the frame of his posthumously realized 1973 *Amarillo Ramp,* natural and industrial tensions inform *Spiral Jetty.* In 1968, Smithson wrote: "The best sites for 'earth art' are sites that have been disrupted by industry, reckless urbanization or nature's own devastation" that could be "cultivated or recycled as art." Forty years later, when *Spiral Jetty* was threatened by oil drilling, the *New York Times* reported that Smithson "reveled in the juxtaposition of industrialism and beauty, decay and rebirth, rot and permanence." Then–deputy director of Dia Laura Raicovich expressed that he "had chosen his site carefully and loved some things that others might call ugly." Many of the artist's unrealized projects proposed that mining companies should host artist-consultants for land reclamation to synthesize "all of the problems that confront the ecologist and industrialist." While some of his contemporaries (like Helen Mayer Harrison and Newton Harrison, who famously asked "How big is here?") engaged artistic processes to revive ecological systems, Smithson may have preferred dregs of drilling in his viewshed.

Although most equipment was removed from Rozel Point, reminders remain—with rotting pylons, rusting drums and pipes, uncapped wells and natural tar seeps—to be noticed or neglected. Visiting *Spiral Jetty* in winter with few people amplifies the force of its unfurl. Wind envelops a body like a second skin. Blood retreats from your extremities, hands and feet, to gather in your core, to warm your heart, to keep it beating. The otherworldly landscape swallows a body into the reality of its smallness, the gravity of its orbit. On the ground, seemingly isolated, relationships heighten: connecting art to tar to salt to water. Actions recalibrate. Each movement demands presence. Senses awaken beyond sight. Life reduces elementally. Human presence is thrown into relief: as perceptions (de)form around how we (de)value any place.

Each year at Rozel Point, as temperatures rise, the surface melts: natural asphalt bubbles up from underground sludge and creeps across the face of the earth. Summer brings the stink of melting asphalt and penetrating heat of the high desert. The air can pulse with the flight of pelicans—a huffing *wingbeat-wingbeat-wingbeat*—that hushes you in your tracks. Frozen in winter, the tar seeps awaken in summer to remind any visitor to watch where we step, to take a closer look, to question the scale of human acts in relation to the agency of the Earth. Rozel Point resists easy meanings: of what is valued and not, of beauty and ugliness, of purity and ruin, of benefit and loss, of access and distance, of place and displacement. Binaries break down. Fear and hope. Danger and safety. Conserved and wasted. Wounded and healed. In a world characterized by oppositions, Rozel Point reveals the wide gray space between. To walk the mudflats, surrounded by natural death traps in a reputedly dead sea, blurs life across time and place. Separations melt: I to we. Under our skins thrive microbial ecosystems. Sloshing in our ears, biorhythms reverberate as microseisms, humming as a heartbeat or the ancient music of the spheres. In a seemingly timeless yet timestamped landscape, dormancy turns agency. Urgency. Everything is here, there, then and now. In danger of getting stuck in habitual traps, a footprint evokes a human body; a feather conjures a bird. Ugliness beautifies. Darkness meets light, as binary stars, falling into each other's gravity and interdependently orbiting, while constellating with other shifting stars.

When I look at Great Salt Lake, as with *Spiral Jetty*, my sense of center shifts—seeping around islands and shorelines, curving beyond sight. The lake suggests elsewhere, just as natural asphalt seeps into other landscapes, plumbed by oil rigs or paved over by manmade tar. Everything sticks together. As borders elude in the basin, the sun shifts and stings my eyes to water. Wind blows in different directions. It is hard not to think of seas inside us, evolving origins of species in tandem with bodies of water and of land. To trace the shifting lake is to find elsewhere, here.

Weeks speed ahead.

Two months after first visiting the tar seeps, right before I leave Utah, I return to Rozel Point.

It is the end of April, and spring is ablaze with blooms. Tulips, crabapples, daffodils, magnolias. As temperatures swing between seasons, snow streaks the mountain peaks. Due to rain and snow, a previously planned group trip with Jaimi, Greg, and a student researcher has been postponed to today, when the forecast is clear and warm, verging on summer.

Jaimi pulls up to my apartment in a Westminster College van. The student researcher, Kara Kornhauser, is in the front seat. She is a junior from Illinois, a first-generation college student who has received a summer research grant to set up camera traps and study the tar seeps with Greg. Two of her classmates, Chloe from Idaho and Emily from Pennsylvania, sit smiling in the back seat. Greg isn't in the group because he is in Spain teaching a course on natural history collections, but a local artist named Trent Alvey has joined. I sit in the open seat beside her.

As Jaimi drives through the city, we each talk to the person beside us. I converse with Trent, a native of Utah, who supports her husband on an ecological research and conservation organization, Round River Conservation Studies, that regularly takes them across the world. This summer will bring them to Botswana, and she is interested to compare Great Salt Lake with the dry Lake Makgadikgadi in the Kalahari desert.

As we drive north along I-15, suburbs sprawl. Warehouses line the highway and train tracks lie among houses, edged by green grass and budding pink trees, under bright blue skies.

Turning between the front and back seats, Chloe and Emily describe their summer research projects with the GSLI. Each is developing research questions and focusing on a local geography to apply them. While Kara works with the camera traps at Rozel Point, Emily will study extremophiles in Great Salt Lake for bioengineering and neurobiology. Chloe is interested in microinvertebrates in canyon creeks as markers of water quality. Their conversation ranges from coursework to agar art in Petri dishes to culturing bacteria.

"If Chloe ever discovers a bacterium," Jaimi jokes, "it will be named *E. Chloe* instead of *E. Coli.*"

Jaimi mentions that she will be going with the GSLI director Bonnie Baxter to the Dia Art Foundation in New York in October to talk about Great Salt Lake beyond *Spiral Jetty.* The students animatedly describe their efforts to highlight women scientists who study Great Salt Lake, affectionately called "Ladies of the Lake."

We pass Ogden, Brighton City, the Bear River Bird Refuge, occasionally smelling the stink. The Wasatch Mountains retreat in the rearview mirror as hills roll down to marshlands. Near the turnoff to Promontory a NASA sign reads "POWER TO EXPLORE" above smaller text that reads "Space Launch System." It is strange to see, again, these rural and cosmic fusions. Dirt fire roads crisscross hills where cows graze. The billboard about "SAFETY FIRST—ALWAYS" comes after a small sign for "Orbital ATK Systems."

At the Golden Spike National Historic Site, the parking lot is full of cars. The visitor center is open, and a few people amble around the store. It is my first time inside. Racks of books about trains and regional history line the shelves, alongside tchotchkes of faux golden spikes, chocolate lumps of coal, and a sign with hobo symbols: *Go this way, Keep quiet, Good place to catch a train.* Parallel lines indicate *The sky's the limit.* Outside in the back along the train tracks, a school group listens to a ranger.

The park store manager tells us about the upcoming sesquicentennial celebration planned for May 10, 2019. Up to twenty thousand people are expected: a mix of train aficionados, history buffs, and descendants of railroad workers, commemorating their ancestors who gave their lives for this transcontinental connection.

Jaimi introduces herself and says that GSLI will be responsible for the road beyond the historic site to *Spiral Jetty* if visitors choose to visit. She doesn't know how the single rutted road will handle that level of traffic, how the lake will respond to crowds, whether anyone will get caught in the seeps. She hopes that the Park Service and GSLI can coordinate as the date approaches.

The store manager nods, and they exchange contact information. "*Jetty* people have a specific look about them," he says, mentioning something about "black leather pants." None of us are wearing black leather pants.

The schoolkids swarm into the store. He looks at them, then at us, and apologizes that he needs to attend to the growing line of customers.

Back in the van, we get sandwiches from the cooler and eat during the ride. The pavement ends and transitions to a bumpy dirt road.

"Watch for burrowing owls," Jaimi says, telling us to look for holes in the ground between lumped sagebrush.

The van rides rougher than February, due to the lack of shock absorbers. It is hard to hear each other over the rows of seats.

Kara spies a burrowing owl, then a curlew with its curved chopstick beak.

Binoculars get passed around.

"Have you visited *Spiral Jetty* before?" I ask our group.

Trent has visited. The students have not.

"I'm expecting it to jump out," Emily says. "I keep looking ahead and trying to figure out where it will be."

As she describes her expectations, *Spiral Jetty* sounds like an extraterrestrial crop circle.

When we mention that it is manmade, she looks a bit disappointed. We refrain from saying more before the newcomers encounter the site firsthand.

The front seat hums as Jaimi and Kara discuss the plan for Kara's project. Today, they will work on the road counter and send the rest of us off to explore *Spiral Jetty* before reconvening to set the camera traps. Jaimi encourages Kara to come back another day for five hours of "ground truthing," checking that the computerized road counter accurately reflects the numbers of visitors. She warns us about no-see-ums, adding that she brought bug spray, so we don't get spotted with itchy bites.

The ride feels entirely different than in February. Unlike that ethereal chilly blur, today the sun beams through clear blue sky. In the backseat, I am dizzy from the ride, catching only snatches of conversation. As we approach the lake, pelicans appear in thick flocks. The lake already looks higher, coming closer to the shore.

Jaimi stops the van at the striped rock.

We cover ourselves with sunscreen and bug spray. Jaimi and Kara start digging up the road counter. Emily and Chloe walk ahead. Trent and I follow at a distance.

Cars already fill the makeshift parking lot. There are crowds, at least two dozen people. It seems like a different place from my visits in October and February.

No-see-ums nip our skin.

The temperature feels warm, almost stifling.

I had planned to climb the bluff to repeat the aerial view, to see the shifted light and reconnect with the expansive scope. But it feels wrong to cut off from the group and isolate myself from the mix. Emily and Chloe have gone straight to *Spiral Jetty,* following its coil. Trent and I leave the road and start to descend the rocks in their direction.

As soon as we drop to the shore, a slight breeze rises. The bugs subside.

As we step over and around the salt-encrusted black basalt boulders, *Spiral Jetty* unfurls but doesn't hold my attention. I glance back at the bluff but am more distracted by the mudflats, the risen lake, the shoreline edged with bright orange. Orange, not pink nor muted, almost florescent. The color flares and lures me off course. Steps away, the salt water seems within reach. I never got this close in February. The lake was too far.

My feet veer off the rocks and onto the mudflats, approaching the lake's edge.

The luminous burnt swirls remind me of tar seeps on fire.

I touch the salt water for the first time.

My skin tinges orange.

Trent follows. Within minutes, Emily and Chloe leave the spiral and start approaching. More people start coming to the lake's edge.

"There are the halophiles," Chloe says.

A family walks past us and steps into the lake. Their kids play in the shallow waters. Great Salt Lake is very shallow, averaging only around fourteen feet. You can walk out for what seems like miles.

"Is it a lake or an ocean?" the boy asks.

"A lake," his father says.

The children play some more.

"I just found a really cool rock in the ocean!" the same boy hollers.

A lone tree, almost leafless with a skinny spine, grows from the water.

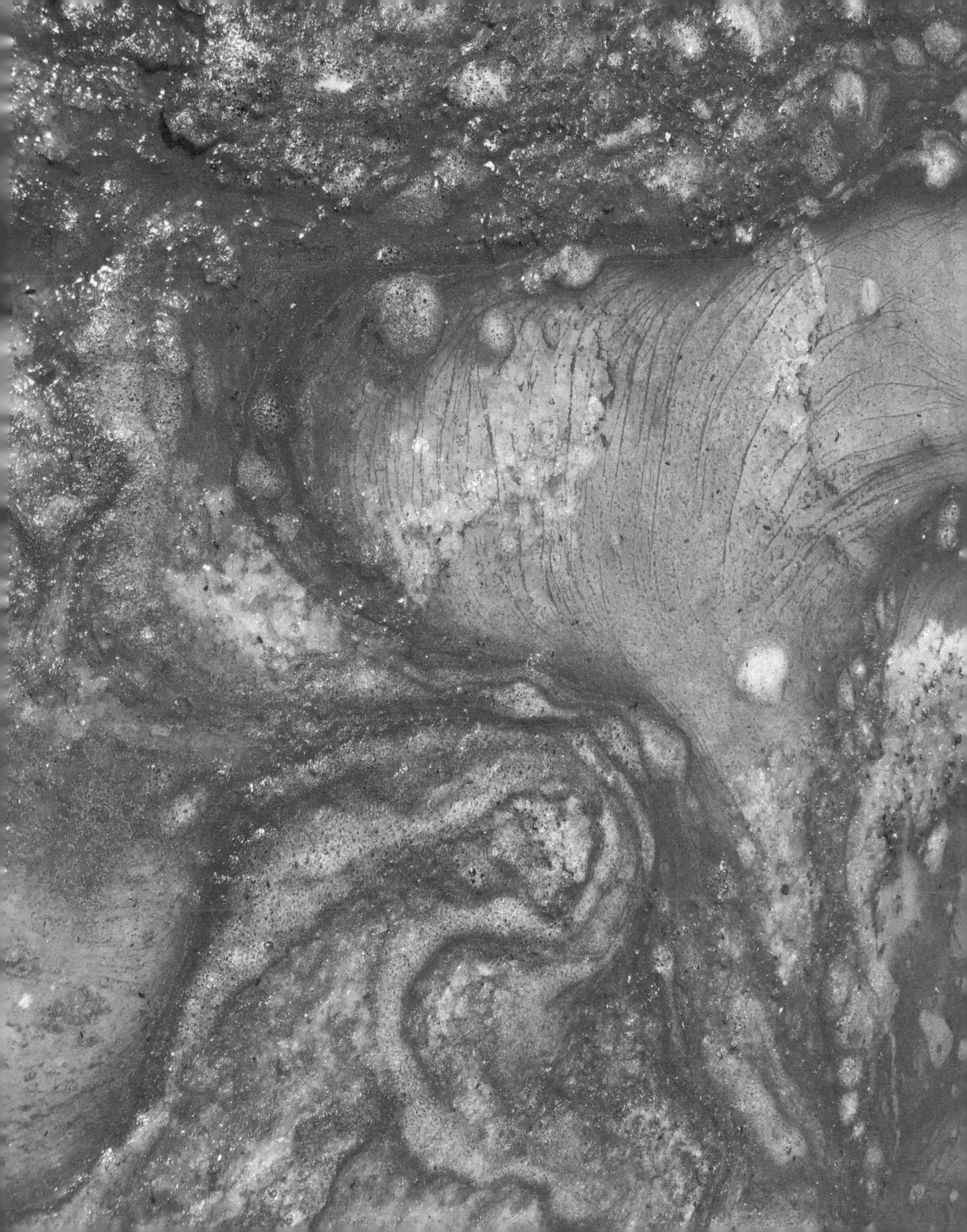

Pelicans fly in flocked formation over our heads, like a traffic jam, alternately relieving their nests to make the long flight to feed at the Bear River Refuge. They whistle through the air.

The lake feels alive, unlike its hibernating state in February.

Birds, bacteria, bugs.

Emily paints her arms orange.

More people wade into the lake, including another family.

"Did somebody die in here?" another boy yells.

We amble around the water's luminous edge. I photograph the swirling orange patterns. Jaimi and Kara join us. We run into one of my colleagues from the university who has lived in Salt Lake City for fourteen years but has never visited *Spiral Jetty*, coming today because she is researching the railroad at Promontory. More people come and go. The lake seems to be luring more and more visitors from the shore.

Pelicans keep flying overhead in V-shaped droves.

Jaimi says we need to start setting the camera traps.

While she and Kara drive the van to park closer to the pylons, the rest of our group walks along the flats to the seeps.

As we walk, the temperature gets hotter. The sun beats without shade. Most of us left our water bottles in the van. Everything feels different: the people, the salt, clear blue skies, the heat. More cars drive in and out. As we walk farther from *Spiral Jetty* toward the seeps, the white dot of the Westminster College van follows the road and turns down the incline to park. Two ant-sized figures get out of the van and grow as they approach.

Kara pulls a wagon, full of equipment. Its wheels leave parallel tracks across the mudflats.

As the six of us walk to look for the tar seeps, we pass rusting oil drums, rotting pylons, and small black puddles of hard natural asphalt.

All of a sudden, globules of tar glisten. The black takes on a deep brown glow, almost auburn, lit from within. The edges are melting. The seeps are awaking.

We pass a few small seeps before finding the one that Greg had called the “pelican death assemblage.” Kara parks the wagon and starts examining the feathers, bones, tar. I walk around taking photographs, looking for similarities and differences from February. Trent joins me, and we marvel at the patterns.

Kara mounts the camera trap and calibrates its settings. Out of the wagon she pulls a laminated sign and affixes it to the metal stand. It reads: Please Don’t Steal My Cameras!!! I am studying the Rozel Tar Seeps in order to learn more about the animals that die here, the seeps themselves, and to provide the public with information about the possible dangers of the seeps!

A seagull flies above us, sweeping back and forth. It flies lower and lower, at times flying over the seep.

Watching the gull drop closer to the warm tar, almost hovering over the bones, I want to warn it, try to scare it away. I don’t want *this* bird to be the one that triggers the wildlife camera, through motion and temperature, right before our eyes. Or even down the line, when a student or researcher might watch on film as *this* bird dies in slow motion. A *not-here* bird feels less real, more abstract, existing only as bones and stones. It’s hard to play time backward to see the stones and bones take on feathers, fluttering alive, winging.

While non-invasive, the camera traps ultimately will chronicle the process of dying through natural entrapment: frame by frame, webbed feet and wings will struggle to get free. That is the point: to represent life at this moment, future fossils in the making, to know that what is alive now will be dead on the continuum between dust and cell and star.

I am reminded of the dog who was trapped, and the photos that showed it wedging deeper and deeper, harder and harder to remove, as the seep swallowed its attempts to escape. Like a mouth on the earth, the seep lies waiting to eat. To be saved, the dog needed twenty strangers who happened to be in the right place at the right time.

I am in this place, at this time, and want to scare this seagull away. I want to scare it to save it.

I walk toward it, catching its eye.

The gull and I look at one another.

I take a step forward.

It steps back.

We begin our strange dance.

It walks faster as I follow.

Faster, faster. Away, away.

It rises, flying over the seeps, and lands again. It walks toward the seeps and bones, oblivious to their sticky threat.

I step toward it.

Again, we tango.

Faster, faster; away, away.

The gull rises and swoops, again and again, landing inches from the seep.

I feel as if watching a dinosaur, a surviving relative, here long before us, with ancient knowledge stored deep in its bones.

I want to scare it away, but then stop. It's not my place.

The gull swoops and lands inches from the seep.

"Let's find the owls," Jaimi says.

I look back at the group. Kara's wagon is packed. Her camera trap is set and signed. We all turn and walk in the direction opposite of *Spiral Jetty.*

The wagon leaves parallel tracks across the flats.

The seagull rises and follows us, flying overhead. I watch it out of the corner of my eye.

Above us, pelicans keep commuting in flocks from Gunnison Island to the Bear River Migratory Bird Refuge.

The flats stretch far in front of us.

Later, I will share a small part of Rozel Point—with its tar seeps, owls, pelicans, and gulls—with some environmental scientists from around the globe who convene around increasing threats to biodiversity and diversity in the climate crisis. Among our conversations, we will question the carbon footprint of our collective travels, as if the purchase of emissions could offset our impact. For another meeting, our departure from Milan's airport will coincide with the day that Italy becomes the new epicenter of a global pandemic, which soon spreads through the United States and across the world. In a few months, my native California will blaze from more than eleven thousand lightning strikes, engulfing much of the state in flames and smoke. Later, I plan to move to the Southwest and widen my sense of the West, with all the permutations of that complex place.

Here and now, at Rozel Point, the landscape looks much dryer than in February. Gypsum appears everywhere: large and layered wedges of it. Crystals glint through the sandy surface, easy to dig out of the dry mud. I dig up a chunk, noticing its strata, knowing that if I run it underwater, the layers will splinter into finer and finer clear layers, until they become slivers of their former selves.

Digging up gypsum with Trent, we fall behind from Jaimi and the students. My pockets fill. When we stand up, the wagon is far in the distance, and its wheels have left longer parallel tracks across the flats.

The gull is gone.

Trying to catch up, Trent and I get distracted by a cluster of rotting pylons like a forest of delimbed trees. There are holes where nails used to be. One hole creates a peephole through which I view Great Salt Lake as a miniature world.

There are no people out here. Very few people come to the tar seeps, mainly by accident when visiting *Spiral Jetty* down the shoreline. Far in the distance, the parking lot remains full. Each departing car looks like an ant that has lost its colony. Out on the flats, there's little breeze but no bugs.

We catch up. In the limbo that follows, we get lost looking for the remains of the barn owls. As the afternoon wears on, it gets hotter. We get thirstier. The wagon tracks meander, then stop.

Jaimi and I are the only ones who have visited this seep, and she asks the group to wait while we search.

The two of us go ahead in different directions. Long ago in a college wilderness first aid course, I was taught to find a lost hiker by spiraling out from the spot

where she went missing. Here, trying to find a tar seep hiding in plain sight, we overshoot the spot, wandering too close, too far. My attempted spiral jags and is jettisoned. We keep getting lost. The lake level has changed; the light glares.

"I found it," Jaimi calls out, finally.

I walk over to check the seep.

It looks utterly different than in February. The big black star has melted.

"See the four owls?" she asks.

The owl bones have yellowed to sepia, not black and white, but brown and blurred. There are four distinct spots of matted feathers and bones. The seep looks muted, washed out like constellations above a polluted city sky. Still, this is it.

We call and wave to the group.

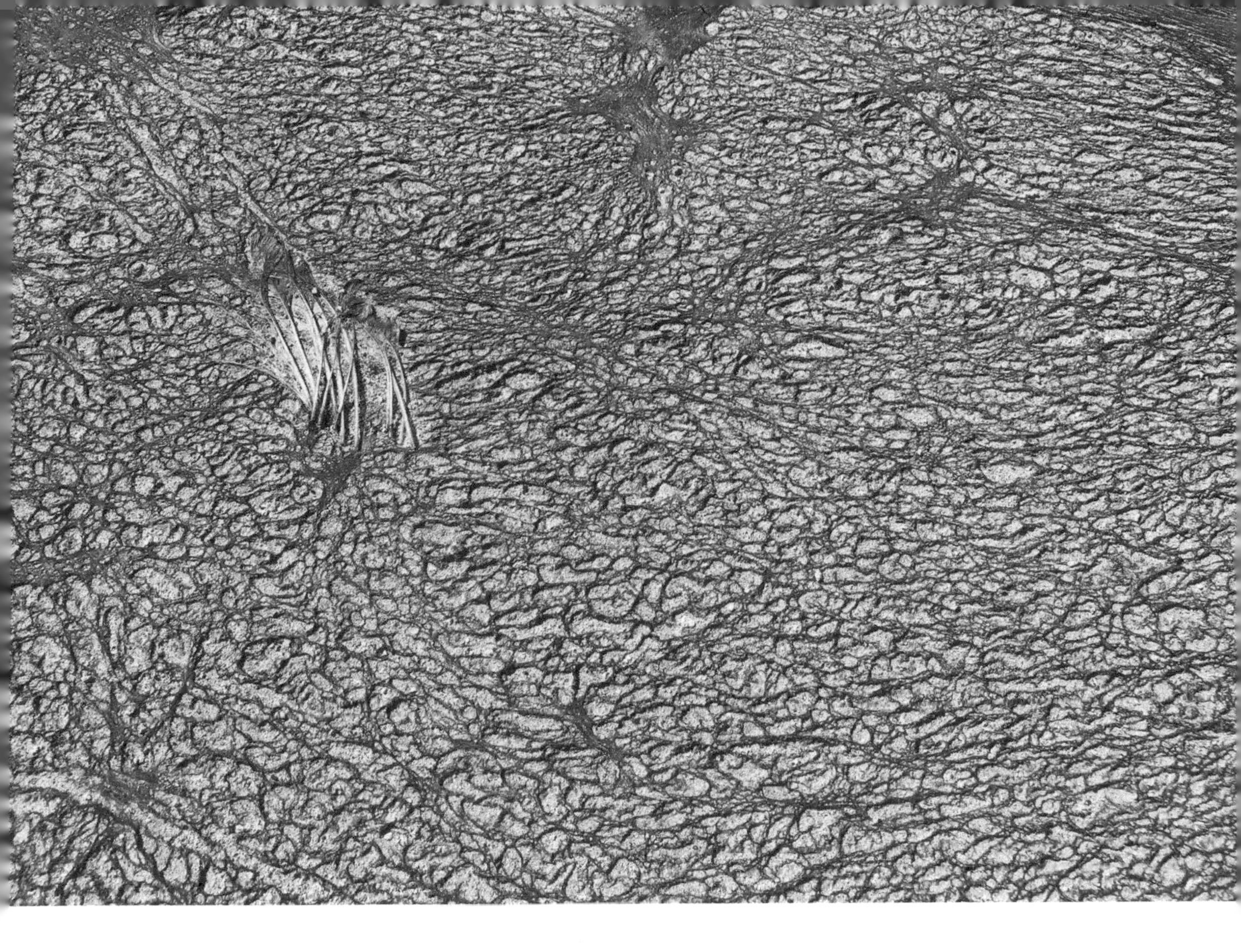

I start walking around the seep. Two months have passed, yet the star has melted almost beyond recognition. It is, yet is not, what I remembered. I am shocked at the difference. I step closer to the edge to take a photo.

The sand beneath my feet starts to sink.

I look down as black, glinting edges outline my shoes.

Quickly, I lift my legs and step back.

Black sticky threads like gum stretch from my soles. My shoes are edged in tar. There is a subtle stink.

I look back at the sand-covered seep. As fast as I sank, the sand has swallowed my footprints back up again, leaving only a faint outline of where I stood. The deceptive sandy surface appears to be waiting for the next unsuspecting wayfarer.

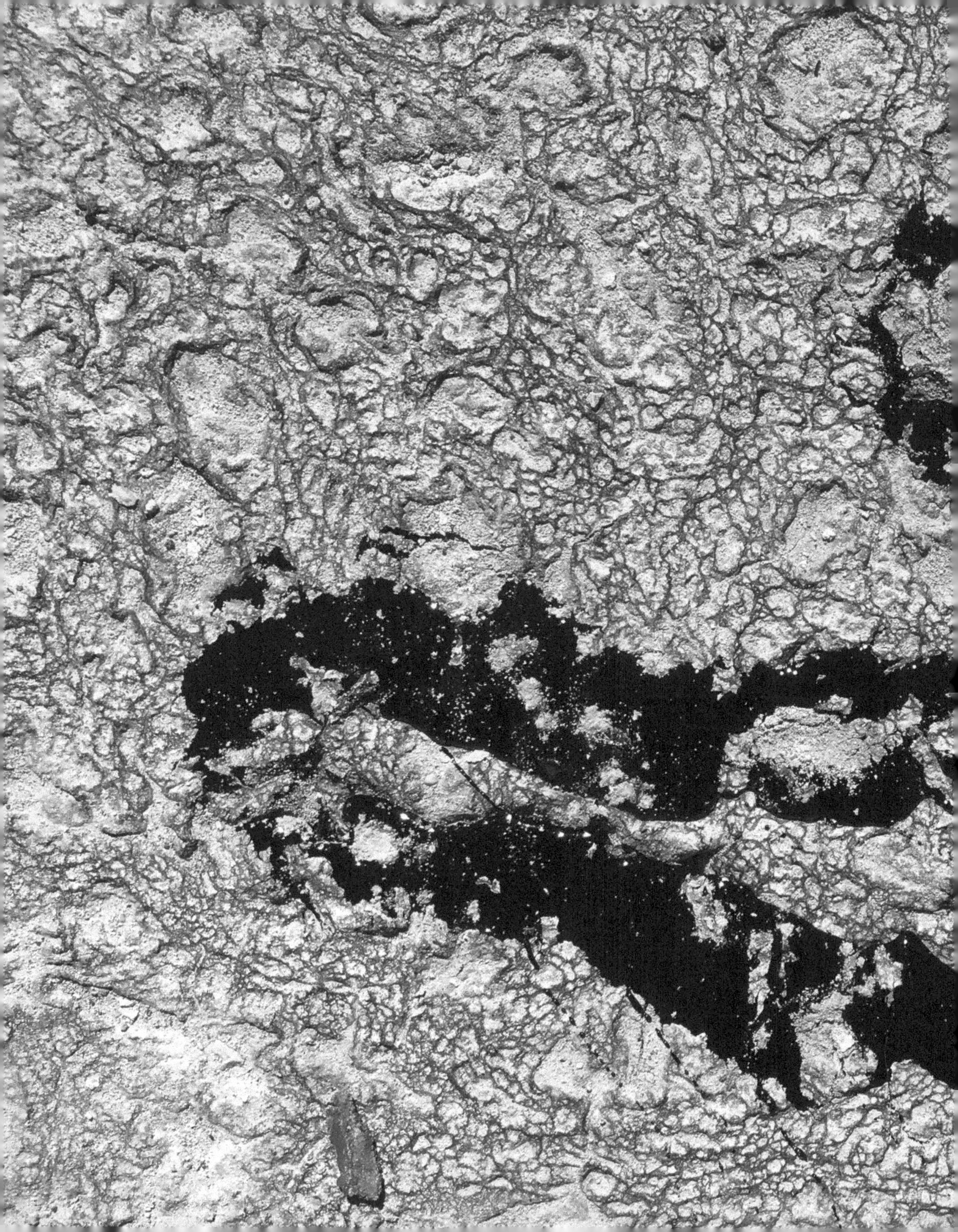

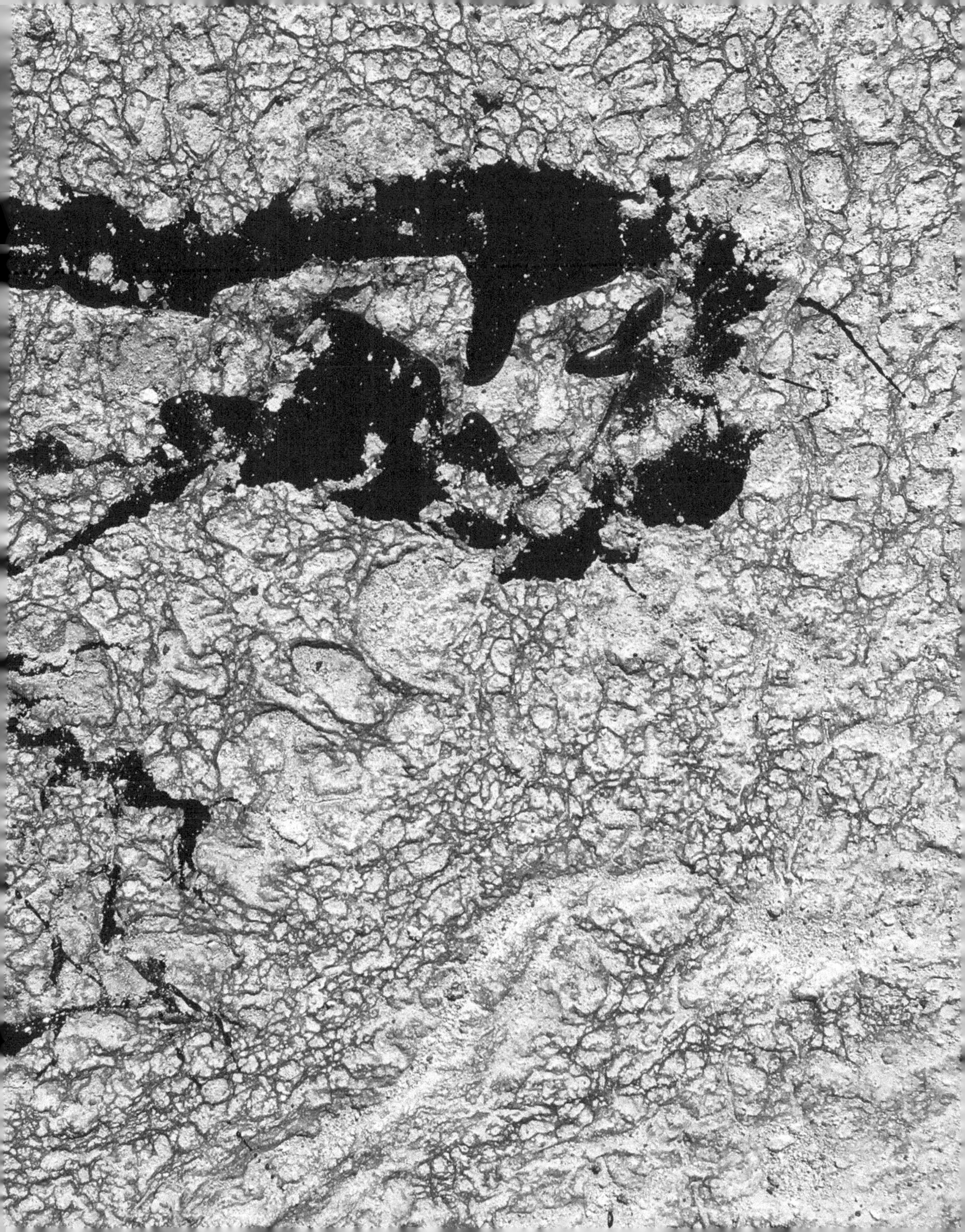

"Watch out for the seeps!" I call and warn the others not to get too close. "They're awake." I am surprised that I fell for the trap.

Only a faint trace shows that I was ever there.

Jaimi walks over to see my disappearing footprints.

The others stay back from the edge, walking and talking, as Kara starts to set up the camera trap.

I start circling the edges again, overhearing Trent and Emily talk about a possible internship in Botswana.

Kara mounts another camera trap, calibrates the settings, bolsters the stand, and adds her sign. She plans to return to the seeps in a few weeks with Greg and a retired specialist from the La Brea Tar Pits. When she announces that she's done, it feels momentous.

The sun beats down. We applaud Kara's efforts and start heading back to the shoreline.

Trekking over the flats, I see a cluster of colored stones, unlike anything I have seen today or in my previous visits. Most clusters on the flats have been minerals: gypsum, salt, calcium carbonate. This bunch yields small round stones—reds, greens, grays, a surprising medley of colors—and I scramble to take one last photo. My hands are crusted with sandy clay. My pockets are full of gypsum. I am overheated, thirsty, clumsy. Struggling to grip the case, I drop my phone.

It falls and cracks on the rocks.

My companions turn at the sound, gasping at my minor mishap.

"Did it break?" Trent asks.

I pick up the phone. Lines spray across its glass face. If not contained, the cracks would spread beyond its frame.

My face crumples. This seems a sign to go.

We walk through the same pylons as in February. Emily now pulls the wagon, relieving Kara, leaving wheel ruts. Someone could easily follow our tracks.

Kara asks to take a group photo and sets up a travel tripod on the ground. She looks through the viewfinder and jokes that it looks like the album cover shot for a band. We photograph the "Ladies of the Lake."

Back at the van, no-see-ums start biting us. We swat and try to keep them from getting inside the van, as we unpack food for the ride. Water bottles and air conditioning feel like delirious luxuries.

As the van bumps along the road, I turn on my phone to see if it still works, to see if the images survived. Opening my photo stream, I start scrolling through frames. The colors seem more muted than I hoped, but the afternoon replays: the orange halophiles, the tar seeps, the camera traps, Kara's laminated sign, disintegrating feathers and bones, the melted star that no longer appears like a star. Frame by frame, I scroll through the hours of the day, distilled to a few minutes. The iPhone's "live" feature was accidentally on, so some shots flutter.

Then I see the gull. Shot by shot, it lands and rises. Its wings spread to fly, right over the sticky pile of bones. The bird hovers, then sweeps lower, and lower. My stomach drops. I want to scare it away.

I wonder where it is now.

I wonder what is happening, now, at the seep.

Again and again, I want to warn the gull, to scare it away, but realize it's not my place. I am the one out of place here.

Like a scare eye floating over a national border, or the thoughtful watch of a hawk, or my great-uncle Fritz viewing the world as an osprey, Great Salt Lake stares out at me from the face of the planet.

The gull flies through my photographs, hovering and landing, swooping and flying.

A peephole appears in a wooden pylon.

Wagon tracks roll away in the sand.

My feet sink in the tar.

The sand covers my tracks as if I have never been here.

I scroll back and see the gull rise again.

The bird swoops and lands inches from the seep. Through the camera, it looks up at me and I look at it, as we eye our fusing future.

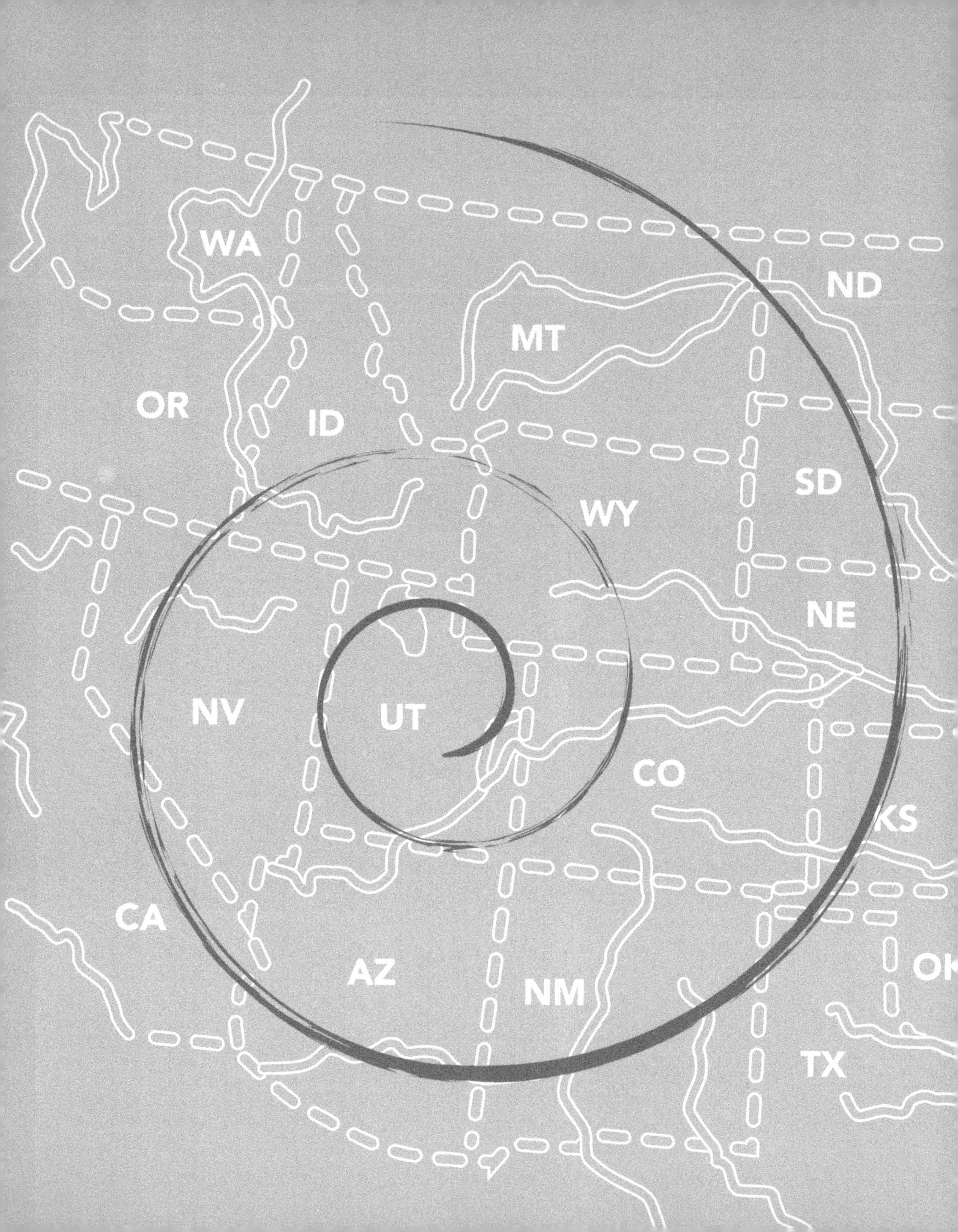
WA
ND
MT
OR
ID
SD
WY
NE
NV
UT
CO
KS
CA
AZ
NM
TX

NORTH AMERICA

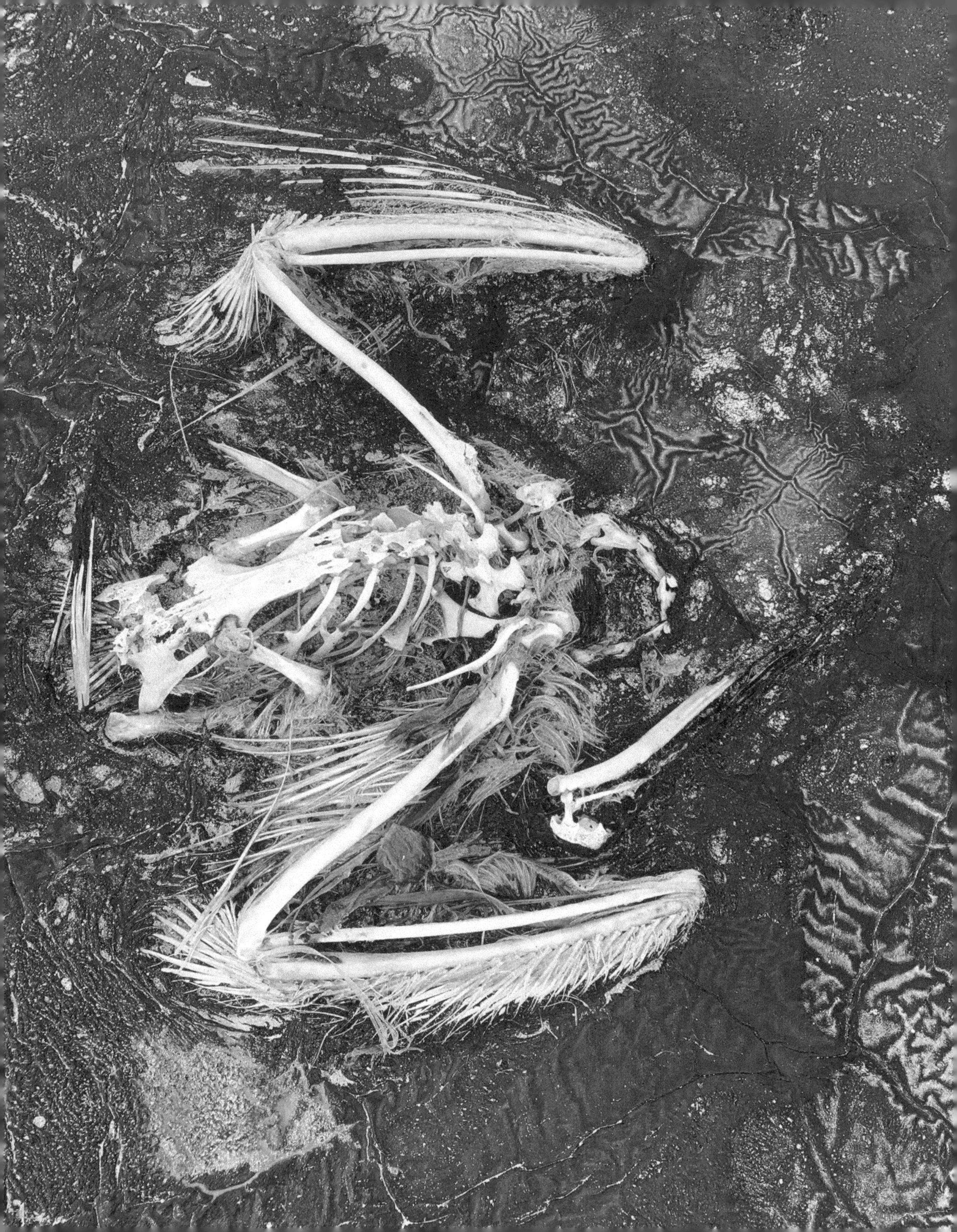

III.

Staying with the trouble requires learning to be truly present,
not as a vanishing pivot between awful or edenic pasts
and apocalyptic or salvific futures, but as mortal critters
entwined in myriad unfinished configurations
of places, times, matters . . .

—Donna Haraway, *Staying with the Trouble*

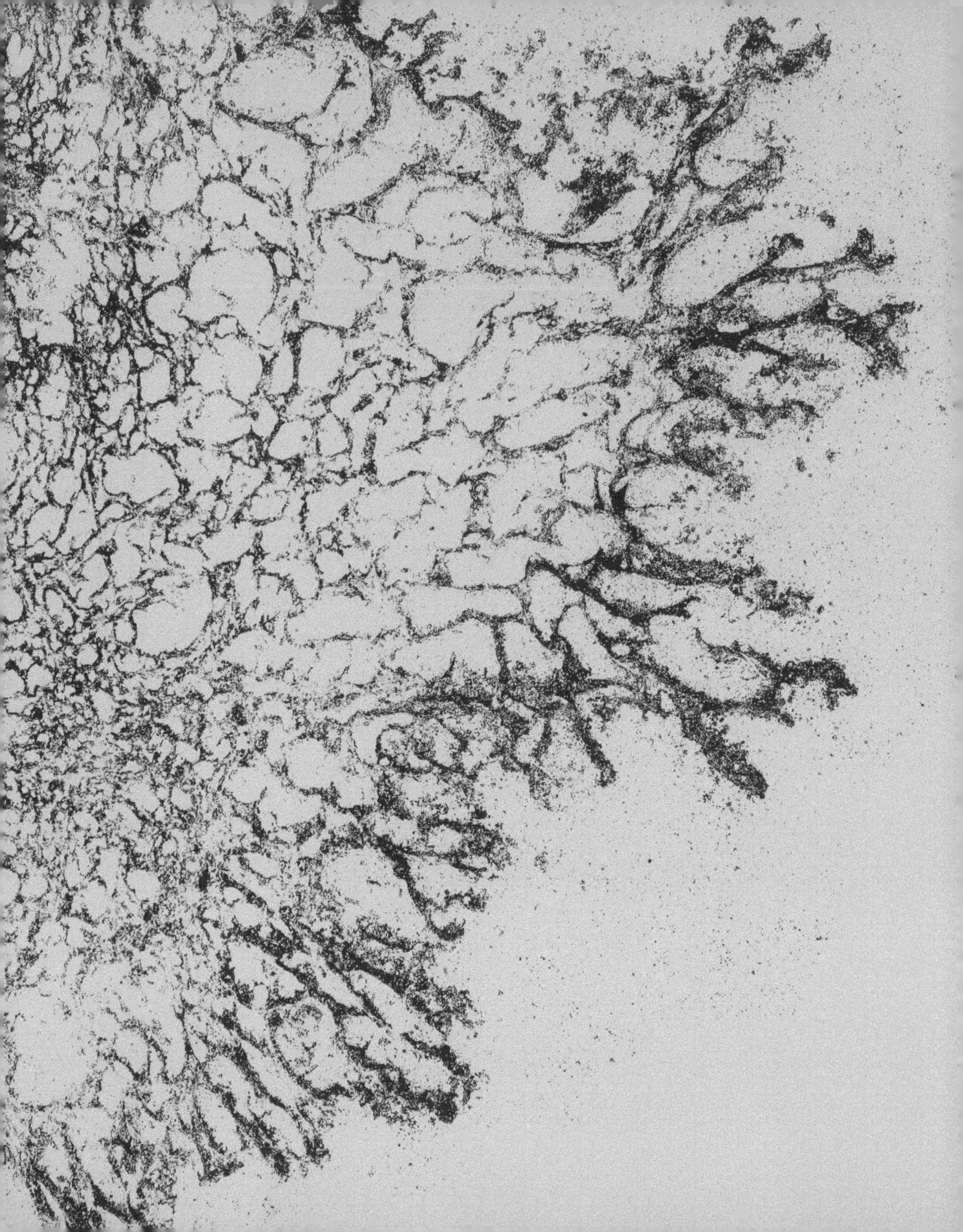

FOSSILS
IN THE
MAKING

Unspiraling

My impressions of tar seeps spiral together over 2017, 2018, 2019. In the end, my return to Rozel Point is repeatedly delayed by a wet winter. Today is my last feasible visit before leaving Utah. It is late March. Rain pours. Visibility is low. On this blustery day, our group drives in two vans. Jaimi and Greg bring one group of students. My van is driven by Kara, who will soon graduate, and is now the lead student researcher for the tar seeps project. The van repeatedly hydroplanes on the highway. We slow to a snail's pace near lanes that are blocked, given a massive sinkhole in the asphalt.

> *asphalt*, according to the *Oxford English Dictionary,* is "1. A bituminous substance, found in many parts of the world, a smooth, hard, brittle, black or brownish-black resinous mineral, consisting of a mixture of different hydrocarbons;

called also *mineral pitch, Jews' pitch,* and in the Old Testament *'slime'*. 2. A composition made by mixing bitumen, pitch, and sand, or manufactured from natural bituminous limestones, used to pave streets and walks, to line cisterns, etc."

At Promontory Point, already thick with swirling snow, our group arrives first. The manager advises us to turn back. The dirt roads are bad; her colleague tried to drive that morning and slid off the road. We'll likely get stuck. "If you decide to risk the trip," she adds, "call someone to tell them your plans. That way if you don't return by a certain time, the sheriff can come out looking for you."

When I go looking for seeps, its variants drip from *sipe,* from Middle English *sipen*, from Old English *sipian*, from *sīpōnan, sīpanan* (compare Middle Dutch *sīpen* 'to drip,' archaic German *seifen* 'to trickle blood'), from *seib* 'to pour out, drip, trickle' (compare Latin *sēbum* 'suet, tallow,' ancient Greek *εἴβω* 'to drop, drip'). *Seep* evokes *see, seem, sip, say, sea, seagull, seal, seed, sleep, slick, slime, slump, slurp, spill, stick, swell,* and other sounds that *slip* through my ears.

. . . abandoned, accident, aerial, apocalyptic, asphalt . . .

After Greg and Jaimi arrive at Promontory, we discuss our options: to proceed or turn back. The agenda is to check the tar seeps and a recently capped oil well, to survey extant camera traps, to set up new ones and tag dead pelicans digitally through "The Collector": an ArcGIS app for capturing field data. We watch fast-falling snow through the window. The usually intrepid Greg admits, "Discretion is the better part of valor." Kara is anxious to proceed but reluctantly adds, "There's nothing wrong with saying no." This is her last possible trip before graduation. "I'll go first," Jaimi says, swaying the vote, "so if someone gets stuck, it will be me."

Tar seeps, also called *oil seeps* or *petroleum seeps,* creep up from tectonic fractures. They are surface versions of tar pits like Rancho La Brea in Los Angeles, which famously preserved mammoths, sloths, and saber-toothed cats. With tar volcanoes and tar balls, tar pits and seeps bubble up in Southern California (underground and off the coast, with tar cones and tar whips) and around the world—including Peru, Ecuador, Trinidad and Tobago, Venezuela, Azerbaijan, and beyond. Where tar bubbles up you usually find oil and gas extraction or, at least, attempts at drilling.

The ride through snow blurs our windshield as we head over Promontory Pass. Flakes swirl over the road, filling ditches and coating hills. Kara wonders if the seeps will be covered and inaccessible. Renae, a student who has processed thousands of photos from the camera traps and tagged them with metadata, voices what we're all thinking: "I hope we don't get stuck." The van's mixtape scrolls through songs to tracks from Paul Simon's *Graceland*. "Slip-sliding away . . . Slip-sliding . . ." "Let's change the song," says Kara, reaching for the dashboard as we nervously laugh.

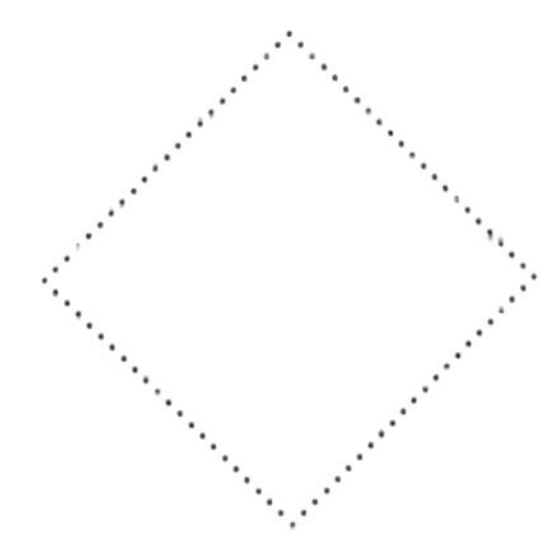

"About one mile north of the oil seeps I selected my site," wrote Robert Smithson about his chosen location for *Spiral Jetty* in 1970. "A series of seeps of heavy black oil more like asphalt occur just south of Rozel Point. For forty or more years people have tried to get oil out of this natural tar pool. Pumps coated with black stickiness rusted in the corrosive salt air. A hut mounted on pilings could have been the habitation of 'the missing link,'" he added, with "evidence of a succession of man-made systems mired in abandoned hopes." After he and Nancy Holt first visited Great Salt Lake, Smithson wrote, "We followed roads that glided away into dead ends. Sandy slopes turned into viscous masses of perception

. . . bacteria, basalt, basin, bitumen, black, black gold, brea (Spanish for tar), bubble, burial, burning water, byproduct . . .

. . . an expanse of salt flats bordered the lake, and caught in its sediments were countless bits of wreckage." Shortly after the creation of *Spiral Jetty,* lake levels rose and covered the artwork for almost three decades.

"We just went through a mountain pass," Kara says as we descend into ranchland. Snow has reverted to rain. "Those may look like hills, but they're mountains." This is her tenth trip to the seeps in under a year. She started a few months after my first trip and now knows the route well. Until Greg and Jaimi started collaborating on this project, the tar seeps at Rozel Point were essentially ignored, not studied except by oil companies and Robert Smithson. Wet windows blur fenced fields of sagebrush, dotted by black cows. A few cows stand in the road. "They're fenced out," Renae says, then pauses and laughs. "Or maybe we're fenced in."

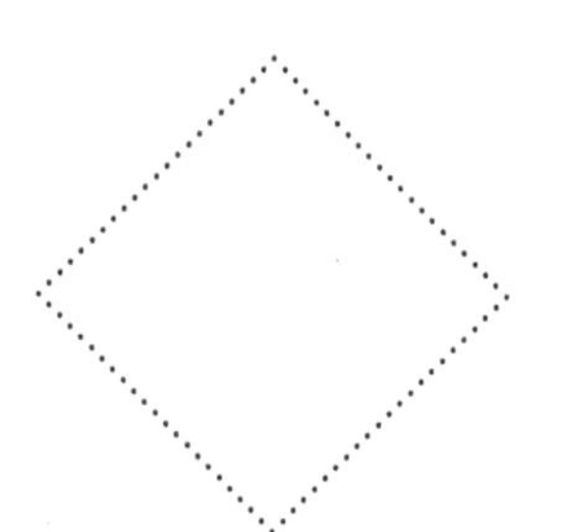

Tar sticks things together. Humans have used natural asphalt or bitumen for centuries dating back to c. 40,000 BC. Across cultures, tar has caulked and mortared walls, waterproofed boats and roofs, coated flint implements, lubricated equipment, protected rigging on ships, branded sheep, treated wounds, and generally glued things together. From constructing the towers of Babylon, to embalming Egyptian mummies (*mummy* is derived from the Arab *mūmiyyah,* meaning bitumen), to filling gaps in Chumash canoes, tar is associated also with providing light: from oil lamps to early photography.

As we approach the bay, the lake appears surprisingly high. Fog, mist, and rain blur the watery horizon as if a mirage. The mudflats pool gray and glistening, lacking snow. We try to discern the tar seeps, eager to see one recent manifestation that we have observed only through drone photographs—a plugged spill, the

. . . camera trap, cap, carbon dioxide, caulk, causeway, charcoal, combustion, crude oil, . . .

retreat of a seep (the "pelican death assemblage")—before and after the capping of an old oil well. Down the shore, the sky seems impossibly clear, streaking blue and white. Compared to where we have come from, the landscape opens as a lens letting in light.

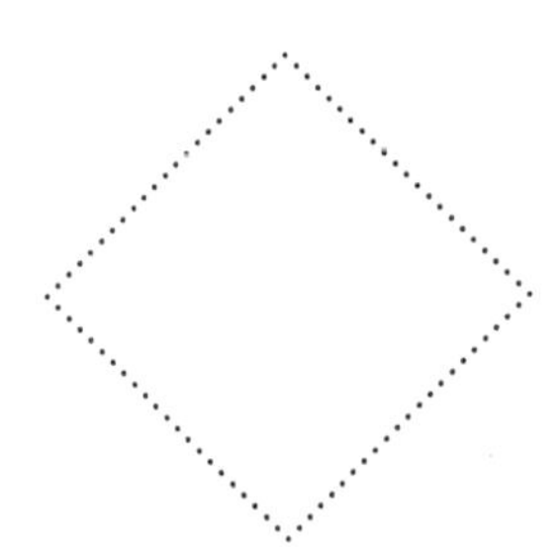

Camera obscuras temporarily captured images that did not last. Around 1826 a French amateur inventor named Joseph Nicéphore Niépce oiled an engraving plate with an asphaltic varnish of bitumen to fix the image with sunlight. After a few hours, the light-exposed solution hardened, as the rest washed away with solvent. He called the result *héliographie,* or "sun writing."

If you visit Rozel Point for the first time, you may first notice the decaying oil jetty and mistake it for *Spiral Jetty.* The oil jetty is straight compared to Smithson's spiral. A jetty is a breakwater; unless the lake level is high, you can walk on both. The jetties invite comparison through proximity. The same goes for the tentacular tar seeps. One thing that I retain from repeated visits to Rozel Point is how these seeps around a spiral can change a sense of time, curling inward and outward, as fragments of birds and viewpoints—yours, mine, ours—can seep through and stick together.

"Did you bring your scuba gear?" Jaimi yells, as a joke, after parking. We lunch in the vans to conserve warmth, then prepare supplies and add layers of clothing. The air is brittle. Standing in the chilly dirt parking lot and checking equipment, Kara says, "This may be the most submerged that we see *Spiral Jetty* in our lifetime." The college group has brought knee-high waterproof boots, but Greg and Steve (a retired paleontologist who drove five hours to join today's visit) only brought walking shoes. My hand-me-down sneakers have black edges from my previous visits to the tar seeps. Walking out to the mudflats, the students splash ahead, as those of us who lack boots watch where we step.

. . . danger, dead sea, death trap, decomposition, deep time, derrick, drain, drill, drip . . .

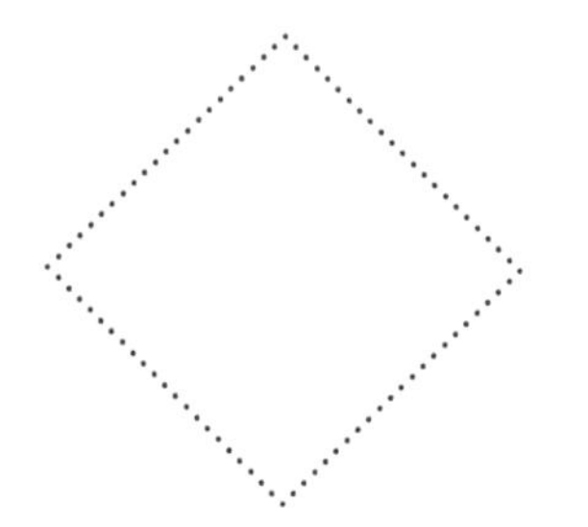

Artists have developed different relationships with land. Some inscribe signatures on the scale of a monument or memorial; others work indirectly or ephemerally over time: tracing the pattern of a walk, leaving elements to decompose, or aiming to direct light. A line made by walking. A trail of rocks. A controlled crack. A volcanic crater turned observatory. A field of lightning rods or bulldozed scar or quartet of concrete tunnels. A border of balloons above a phantom wall. Great Salt Lake has inspired more: a rock jetty as decomposing spiral, a white gallery encircled by color strips, a labyrinth of salt merging with the saline lake.

Whenever I fly above Great Salt Lake, the landscape abstracts into a bird's eye view. Peaks ascend from its silvery spread, curving around cragged bays and islands. The straight causeway cleaves the lake in two, as color fields etched by mineral ponds. Smoke curls from power plants. Mountains ring the horizon with snow. High above, in a commercial plane, I miss details curling around the shore until I recognize features, reorienting me to the Point with its spiral and seeps. I imagine seeing beyond the blur of my aging sight by squinting into the light, as forms come into focus.

As we walk away from the shoreline, Greg explains the source of the seeps in geological terms: how the bluff behind *Spiral Jetty* is a basalt floe, a block of downdropped rock, with sediments atop an older version of Great Salt Lake being converted into raw oil. Trapped in the fractured basalt, the oil forms a reservoir that seeps to the surface. "If you go to Google Earth, you can see the seeps in a line following the fractures." He points out varied lines: fault lines that trend with the road, the Bonneville snowline, and channels along the mudflats, now filled with water where cars get stuck and leave deepening grooves.

. . . Earth, elegy, energy, entrap, entropy, environment, escape, excavation, extinction, extraterrestrial, . . .

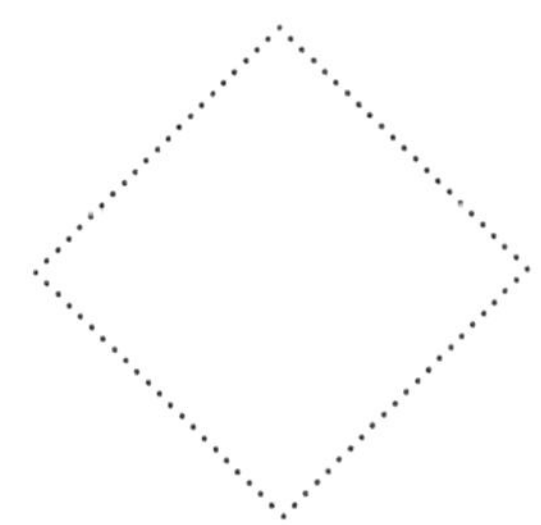

Tar seeps differ from *tar sands* or *oil sands,* like those found in Alberta, Canada. *Tar sands* yield a different kind of seeping from extractions and emissions, where wastewater and tailings spread to marshlands and river systems that spread to ponds to fish to birds who try to land on the seeming water and get fatally coated. In 2008, 1,600 birds died in a single landing. The Alberta tar sands are the sought source of oil for the Keystone XL Pipeline to bring raw oil from Canada to the Gulf Coast to be refined and shipped worldwide. It has been nicknamed *dirty oil* and *tough oil* in the midst of the climate crisis that confronts the world's unsustainable overdependence on oil that has threatened environments and inhabitants, including through *oil spills* in the Gulfs of Alaska and Mexico.

My course involves field guides for a semester-long assignment. Students adopt an unmapped site on campus to return to over months to get lost: observing, listening, questioning, documenting, mapping, counter-mapping, rewriting. *Draw lines,* I urge. *Read between the lines* of sites and sightlines, fault lines, shorelines, pipelines, property lines, meander lines, desire lines, lineated poems and paragraphs. I invite geologists, a conservator, an artist, museum curator, cemetery caretaker, librarian, archaeologist, urban planner, and other practitioners who offer different stories of stones. *Draw lines that brought you here,* I prompt, distributing photocopied maps. *What was here before what is here now?* The political and historical maps divide states and territories by lines, but the geographic map unifies the continent. Beyond coastal outlines, it is hard to orient ourselves, until we see the dot of Great Salt Lake that wordlessly signals: *You Are Here.*

. . . fault, fault line, fault zone, fissure, flow, flypaper, footprints, fossil, frozen, future, . . .

Greg points out the pattern of a pelican getting trapped in tar: stuck feet-first, part by part: splayed wings, tarred feathers, pushing up with its beak, ingesting asphalt. A consistent pattern emerges as pelicans get stuck, splayed, to die in slow motion. "You can see the string of neckbone," Greg says, pointing to the extending neck and lower jaw. A blade of backbone sticks up, fodder for coyotes, left to the elements, while the fossil site gets preserved beneath.

. . . gas, geology, geothermal, Golden Spike, Great Basin, Great Salt Lake, grief, ground, groundwater, gulf, gull, . . .

The pelican was a common symbol in alchemy, vulning its breast while yoked with ideas of sacrifice and resurrection, transformative as the Philosopher's Stone. Pelicans have gotten stuck in tar pits and seeps, even crashing into asphalt roads when summer's heat induces a shimmer. Tar gets stuck with symbolisms, congealing fragments, as associations defile. Consider the tar baby. Tar-and-feathering harks back to the Crusades, also a tactic against British tax collectors in the American Revolution, also used to chase Brigham Young west before he came to Utah. "Tar heels" became an accolade for North Carolina (deriving from tree resin, not petroleum, that gave the name Tar to the river that ferried its barges), where the locally sourced material became eponymous with the worker. Its source degraded from "tar on the heel" as a racial and class slur. Although "tar" has many other meanings (a musical instrument related to a si*tar* or gui*tar*, an acronym for a computer archive, a genetic syndrome, or a feature on Mars), it often is found at a site where someone gets stuck.

"Smithson said nothing about pelicans," says microbiologist Bonnie Baxter, Director of the Great Salt Lake Institute. Her slide echoes: "[Nothing About Pelicans] –Robert Smithson." She looks at the audience and asks, "Why didn't Smithson see them? They are like pterodactyls flying overhead. How can you miss them? When you leave something out, a scientist can draw conclusions." In Salt Lake City, the Spiral Jetty Partnership is presenting their collaborative stewardship of Rozel Point. It is March 2019 after an overnight blizzard. Bonnie talks about her two decades of scientific studies at the lake, where she first floated by canoe over the submerged earthwork to collect specimens. She is joined by curators from the Dia Art Foundation, the Utah Museum of Fine Arts, and a manager from the Division of Forestry, Fire and State Lands. (All are women, and since the blizzard has closed daycares, one curator brings her two young children who play on the floor, enlivening the mood of the room.) Stewardship is not only about the spiral but also the mudflats on which the earthwork rests in Great Salt Lake, in the Great Basin, ever-extending. Smithson circumvented the structure of a museum, intervening in nature to set conditions for the artwork's evolution. As *Spiral Jetty* receives increasing attention onsite and online, the partnership promotes communal care of the site, manages visitor access, documents environmental change, and fosters education: around scientific studies, field tours, children's programs (like art-nature backpacks checked out of the local library), art classes, and coordinated archives. This cooperative model brings together unexpected partners to cultivate care around common ground.

"We have this image of tar as the death trap of the ages," says Greg, as we walk around the seeps with dead birds, "but in terms of the fossil record, tar is trying to protect them. There are lots of implications here for understanding other tar pits and seeps." He is interested in the fossil record; Jaimi is interested in bird migrations; Bonnie is interested in halophiles; others are interested in fly larvae, brine shrimp, microseisms, snowmelt, and other aspects of the lake that are not separate but interrelated. In the field, these accumulations or dissipations teach students to look for patterns, honing their ability to observe and attend to interactions in the natural world of which they are a part. The process isn't something

. . . halophiles, hazard, healing, heartbeat, hope, horizon, hot, humanity, humility, hydrocarbon, . . .

that can be learned from a textbook or overnight. Only by practice. On the mudflats, students engage questions through rubbery tar, sulphuric smells, crystalline gypsum, and saltwater puddles. In knee-high boots, they splash toward water-bound camera traps and gather data. Their arms almost flap. We watch them at a distance as they sink in the water, rising on the rubbery tar, intrepidly sinking and rising again, as they trudge from bird to bird.

As the centennial of the 1918 Migratory Bird Treaty Act, 2018 was named the Year of the Bird. "What would you pay to hear birds singing in the morning?" ask researchers as environmental values grow neglected and commodified. "$2.50? $250? Can we put a price tag on this?" In 2015 the *Oxford Junior Dictionary* replaced natural words with terms of technology, removing a number of birds: *pelican, raven, heron, kingfisher, lark, magpie, starling, thrush, cygnet, wren,* more. Seemingly small acts aggregate alongside government rollbacks of environmental protections including the Endangered Species Act of 1973. In 2019, the United Nations reports that one million species face extinction, and a longitudinal ornithological study reports that, in the past fifty years, North America has already lost three billion birds.

. . . I, igneous, inflammable, inscription, international, intervention, . . .

To visit Nancy Holt's *Sun Tunnels* in the west desert of Utah, I reserve a day with my class in April 2018 and follow Interstate 80 past Great Salt Lake, past the Bonneville Salt Flats and Wendover (where bombers trained), into Nevada and back into Utah, to a highway to dirt roads, to what feels like the middle of nowhere. A rutted road leads to another, winding farther into the desert beyond cell service and amenities. Among dilapidated shacks at Lucin, a metal sign stands riddled with bullet holes around "this quiet spot managed for wildlife," an "oasis in the desert" for "migratory songbirds." Low hills rim the flat, dry horizon. Another dirt road leads to another before the *Sun Tunnels* appear in the distance: four giant concrete telescopes pointing in cardinal directions. The tunnels align with sunbeams from the summer and winter solstices, when visitors camp to watch the sun. Inside, constellated holes open to the sky, edged by black sprays (ricochets of other visitors' bullets). The *Sun Tunnels* are big enough to walk through. "Through the tunnels, parts of the landscape are framed and come into focus," Holt wrote, "stars are cast down to earth."

"There's nothing you can do at that stage," Greg says, remembering what Kara called "the day we watched the pelicans die," a visit that she and Greg made in August 2018. They saw the birds trapped in the tar seeps, still alive. We now stand beside two dead pelicans who lie side-by-side. Looking at the dead birds, both with green metal tags, Greg says, "Here are two animals that obviously got trapped in tar, but there are other ones in between. The tar may not have been the primary reason for their deaths." "Tar or no tar," Jaimi echoes, "they may have died anyway." The tagged specimens join the documentation and build up the database for further study. In the past year at the tar seeps, this team has found dead pelicans as well as gulls, owls, a golden eagle, a snake, and little brown birds that decomposed faster than they could be identified.

. . . jeopardize, jetty, jumbled, juncture, . . .

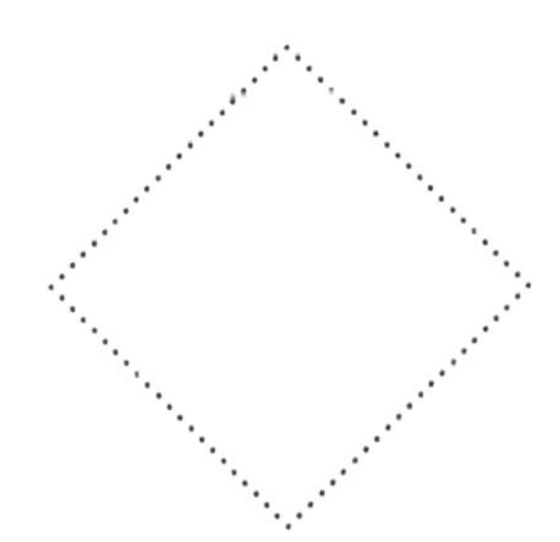

"The Collector" app for "Pelican Mortalities" includes categories for data linked with GPS coordinates to add: *Observer, Date, Species, Other Species Details, Leg Band Number, Wing Tags Present?, Wing Tag Color, Wing Tag Number, Skeleton (State of Carcass), Comments about State of Carcass, Died This Year?, Comments about Died This Year?, Evidence of Scavenging, Notes.* Photographs can be uploaded. Data can be synced in cellular range but not off the grid at the tar seeps.

In the summer between my semesters in Utah, in 2018, I live for a month at an environmental center in Montana's Centennial Valley. Nicknamed "the Serengeti of North America," the valley is part of the Greater Yellowstone Ecosystem with grizzlies, wolves, moose, antelope, sandhill cranes, and more than 240 bird species, including white pelicans. The nearest town is an hour and a half away, so environmental researchers, land managers, and cattle ranchers in the remote region tend to live self-sufficiently and watch out for each other. On Wednesdays, the dispersed community gathers for a potluck to share commitments, and for one dinner, I am invited to share my book project on *Life in the Tar Seeps*. Another evening I see a herd of over two hundred elk. Another dusk brings a thunderstorm with a double rainbow, spattered by sun and hundreds of birds. When I go on a hike with a geologist, she pulls ash from a creek bed and links it to a volcanic peak that once topped Crater Lake in Oregon. Elsewhere in the valley lie the headwaters of the Missouri River that runs to the Gulf of Mexico. Situated over one of the largest known volcanic hotspots on the planet, Centennial Valley is home to the Red Rock Lakes Wildlife Refuge, founded in the 1930s to bring trumpeter swans back from near extinction. For centuries before European colonial homesteaders, Shoshone, Bannock, Nez Perce, and other Native tribes frequented the Valley for fishing and hunting. Cultural and environmental histories layer among living descendants, as researchers from the Nature Conservatory now spend the summer counting birds, as mining ventures scar the state, as forest fires increase. Amid environmental threats, I notice birds anew—remembering how they cross continents and follow migration patterns set in motion eons ago. Over thousands of years, people have associated birdsong with daybreak, a natural stimulant, a sign of safety; when birds stop singing, their silence portends (as *a canary in a coal mine*). Hearing birdsongs is like listening to different languages. Looking at birds, as near descendants of dinosaurs, may be the closest thing we have to looking into the eye of time.

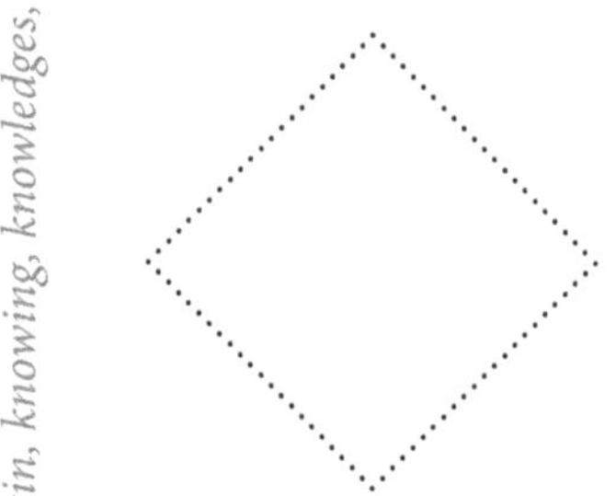

As we wander farther across the mudflats, the puddles spread and deepen into pools, leaving few places to step. Without waterproof boots, Greg, Steve, and I wander the edges, jumping to tar patches like lily pads. We find more dead pelicans, some fresh. My sneakers sink, waterlogged. The tar is not warm enough to smudge: neither melting nor frozen, in an intermediate rubbery state, floating and tangling tentacles. I carefully step to photograph its contortions. From pad to pad, sinking and bobbing. The ground feels buoyant, a floating floor. "This stuff could turn a full-grown adult into a kid!" Steve hollers, laughing. Clusters entwine underwater like kelp, black not green. At first I hesitate to touch the tar, worried it will leave a mark, then pick up a fistful like a small octopus. A student named Johanna finds what appears to be a giant mushroom. The shapes make me marvel, as if creatures are coming into being right before our eyes.

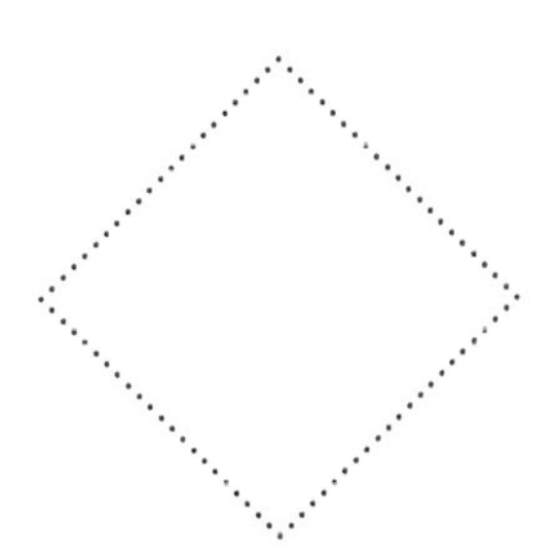

The tale of Great Salt Lake Monster varies across tellings: from a grave robber branded for his crime, to a creature with a crocodile body and head of a horse, to masses of larvae, swimming dolphin-like or spawning a pod of whales. Whirlpools in the lake were thought to drain subterranean channels to the Pacific Ocean or as a river in "communication between the Atlantic and Pacific." Tellings have shifted over time, as if seeping from some truth.

My return to Utah in 2019 coincides with the longest government shutdown in U.S. history. The shutdown is felt coast to coast, far beyond offices in Washington, D.C., to California's Joshua Tree National Park, where vandals destroy the eponymous trees. ATVs make new roads. Furloughed rangers and volunteers are left to clean up human feces. It is estimated that the destroyed trees will take centuries to regrow. Across Utah, there's a problem of visitors disturbing public lands, taking archaeological artifacts, adding graffiti around rock art. Even if a visitor respectfully leaves "no trace" but shares findings on social media, a trail remains for

. . . meander line, methane, microbialite, mineral, mineral pitch, moisture, Mother Earth, muck, mudflat, . . .

others to follow. Despite the Interior's activities that affect 574 federally recognized tribes, no Native American has ever been appointed as Secretary of the Interior (as of this writing). At the University of Utah's law school, former Secretary Sally Jewell gives a public talk saying that public lands are threatened less by climate change than by us humans. She worries that children will increasingly grow up without connections to nature, and paraphrases Wendell Berry's *The Unforeseen Wilderness:* "The earth is not only inherited from the past but borrowed from the future."

As new patterns emerge up close and afar, I continue to photograph the tar seeps. Tentacles seep up beneath the surface to meet those who cross their path. The tar seeps are an interface, a sticky harbinger, a reminder that forces our respect: to watch where we step. The seeps force us to be present. They invite *retreat* in the etymological sense: *to reconsider, to withdraw, to retract, to revoke,* to consider nonintervention amidst our interactions, as we try to leave no trace. Our presence is full of contradictions, driving gas-guzzling cars to be here. As I wander around edges of rubberized tar, I find trash and am reminded of a new type of stone that was discovered in 2006 in Hawai'i that is becoming part of the fossil record: *plastiglomerate.* Melted plastic seeps into pores of rocks or binds together pebbles, shells, and sand that at times bear traces of ropes, nets, lids. I understand why the road to Rozel Point will be closed for the Golden Spike's Sesquicentennial: to keep us from destroying the very thing that we come to see.

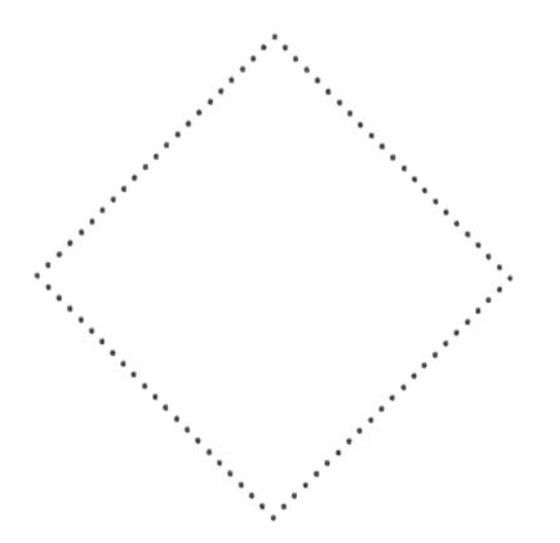

Worldviews evolve through narratives, colliding and coalescing, living as languages. From present-day California, Greg Sarris (Miwok) describes how a team of geologists unearthed whale fossils to match ancestral accounts of a whale in a creek that had been "told from generation to generation for 10,000 years" but were not believed until the bones were carbon-dated. In the Pacific Northwest, Native accounts of phenomena including a massive earthquake and tsunami around 1700

AD (land shaking, everything drowning, water receding then returning, leaving canoes in trees) did not gain credence until a few years ago, when scientists pieced them together with other data from the Pacific Northwest and Japan, crossing centuries and cultures, acknowledging and intermixing languages and worldviews to get a fuller, complex, integrated and diverse picture.

. . . nature, natural gas, natural resources, negation, non-site, North America, . . .

Accompanying a geological team to the Utah–Arizona border in spring of 2019, I listen to seismic vibrations of stones, and recall black tar seeps in contrast to red sandstone and blue water. Since Glen Canyon was dammed in the 1960s, the Colorado River has been cinched as Lake Powell hydro-electrifies the American West. Now the river and lake draw low from drought and demand.

Our motorboat with the National Park Service leaves from Antelope Point Marina. The long ride across the lake bypasses rust-colored towers, buttes, and canyon walls streaked with desert varnish. The open boat ride is cold, but upon docking, we start to bake. After a few bends in the trail, Navajo Mountain rises in snow, framed by the giant red rock span of Rainbow Bridge: one of the longest natural bridges in the world.

Gathered beside the stone, we listen to different narratives about its meaning: by delegates from tribal nations, park rangers, and geologists who share different names for and values of the bridge. I learn why it is important not to walk under this sacred site, why eco-recreation threatens the stone, how its voice is likened to ancestors and may be hurt by the vibration of planes and by sharing seismic recordings with the public, if manipulated. I recall a sign from Arches beside rock art: HEALING IN PROGRESS: PLEASE STAY ON DESIGNATED TRAILS ONLY. Discussions turn to invasive species, tamarisks and quagga mussels, as we walk by blooming cacti and natural shrines and more sacred sites that I do not understand how to perceive.

At sunset, our return by small open boat bypasses vacationers on towering million-dollar houseboats. On the drive back to Page, coal-fired smokestacks of the Navajo Generating Station rise from the horizon. It will soon close permanently, causing concern for local workers. The surreal canyonlands spread around us in all directions, glowing in sunset until they seem mummified as skin, knobbed as knucklebones and backbones, as if the land itself were a supine body on the verge of waking.

. . . oil, oil company, oil field, oil well, ooze, odor, opening, ornithology, oxygen, owl, . . .

Time is running short. I peel off from the group to visit *Spiral Jetty* one last time. Walking along the mudflats, I see gypsum but have no impulse to take it. This morning, we were told not to proceed, when the snow was swirling and slick. Now the sun shines. Barely a breeze. Overhead, a plane flies. Gulls cry. No sign of snow except the high waterline and pooling mudflats. Trash litters the flats: a green plastic cup, crushed bottle glass, an alien painted on a rusting oil drum. Pipes and equipment decompose in black sediment, with sand and rust, like orange-and-black quartz or crystallized microbes. I remember my last trip to Rozel Point, seeing droves of pelicans overhead, and try to imagine Great Salt Lake through pelican eyes: at the convergence of two of the four major migratory bird flyways of North America. Everything spirals together: what we care for and neglect, how fullness emerges from the barren, ugly beauty of Rozel Point with its failed extractions, abandonment and resilience. I feel as if I have traveled to the ends of the Earth and back again, mixing into mud and sand and tar, appreciating this story is less about death than life.

Photography began as a slow process: of sun drawing, of light exposures on plates of metal and glass, believed to steal a subject's soul. Origins of *points de vue* presaged cameras, but the medium forged a new relationship with representation: how we framed landscapes and how we framed ourselves in the world. Over time, exposures quickened. Fast-forward through film. In 1962, Rachel Carson published *Silent Spring* to illuminate the destructive effect of pesticides on the planet. Fast-forward again. In 1972, Charles Stone argued that trees and other natural objects should have legal standing, given the precedent that "the world of the lawyer is peopled with inanimate rights-holders: trusts, corporations, joint ventures, municipalities, nation-states." Leap ahead a few decades to the digital age. In 2015, psychologist Sherry Turkle argued that "we are at a 'Silent Spring' moment in our infatuation with life on screens rather than life in the real world, never wholly in one or the other." Rachel Carson advocated for "conservation"; Turkle promotes "conversation." As a photographic analogy, such quotations are merely "snapshots" where a larger picture remains outside the frame.

Outside the frame:

After I was hit by a car in a crosswalk, doctors told me that as part of recovery, my physical injuries and concussion required "active rest." Physical therapies blended movements like swimming and stretches with speech therapy to exercise my brain to get my body off a walker and curb vertigo, building up my ability to read and write again, without the world faltering or spinning. The first few weeks of "active rest" crucially aided my recovery to a "new normal." When a living body is injured, part of its healing requires rest.

As the planet is over-drilled and overdeveloped, Earth has no chance to rest and heal and repair. Manmade forces repeatedly collide with the land. Natural resources get depleted faster than they can replenish. Scientific studies show how human healing benefits from time in nature, but how does nature benefit from time with humans? Part of the advice of tribal delegates at Rainbow Bridge was to leave a period of rest for the bridge, developing protocols to limit helicopter flights for part of the year, to decrease shaking the stone through manmade vibrations. Living bodies have evolved to repair themselves with astounding ingenuity, even to innovate new or renewed ways of living, but systems can get stuck.

Healing does not mean returning to a past condition. Our bodies grow, age and change, always evolving renewed senses of care. Can we change our relationship with the planet (and with each other) to enable Earth to rest and replenish in renewable ways? What are we willing to give up to make this possible?

The idea of ecological rest underscores how the land is alive, as a kind of breathing, living body. We call Earth a body, with bodies of water and of land. It's questionable where the metaphor starts and ends as an extension of language: a human body standing in for the land, or the land for our body, or both entirely inseparable.

Spiral Jetty appears to float in the shallow lake. The water glints in quiet waves. Pylons turn to people as miraged walkers move upright over the face of the Earth. A family follows their dog across the mudflats. Footprints and pawprints. Clouds shadow islands. I walk from rock to rock, feeling the breeze. A couple walks by in the opposite direction. At the center of the spiral, I stand still. A flock of gulls erupts in a cry, stuttering to silence. In the distance, a car hums along the dirt road. This place is experienced differently over days and years. Taking in the scope, I am reminded of a conversation about musicians who live inside the resonance of sound, returning me along the path that brought me here, attuning to a different dimension.

". . . trace the course of 'absent images' in the blank spaces of the map . . . One is liable to see things in maps that are not there . . . 'the earth's history seems at times like a story recorded in a book each page of which is torn into small pieces . . . the pieces of each page are missing' . . . rushing to and

> from the actual site in Utah. A road that goes forward and backward between things and places that are elsewhere . . . discovered as you go along uncertain trails."
> —Robert Smithson, "The Spiral Jetty"

> "Underground," Greg tells me over coffee in Salt Lake City after our first trip to the tar seeps, "they have found over time that there are certain patterns where the oil gets trapped. But sometimes things happen where the oil comes to the surface, creating the tar pits. That's when you have faulting action. When you have faults or any movement, the seal gets broken. You have the tar pits in Los Angeles—look how many fault zones there are. Where do you see it in South America? On either side of the Andes: where there's tectonic activity. You may have an impervious layer here [at Great Salt Lake] that's trapping all the oil, but if you have something that breaks it apart, the oil can come to the surface. Getting the petroleum to the surface requires some type of tectonic activity—whether mountain uplift like the Wasatch front or slipping like the San Andreas fault—anything that breaks up your packets of rock so there are fractures that a fluid can percolate through. This is also how you get your geothermal springs along the front. The tie-in to tectonic activity is simply a means by which solid continuous layers of rock get broken, which allows this fluid to bubble to the surface, to provide the passageways. It's very holistic—everything ties together."

Back in cell range, Jaimi asks us in the van to sync phones and upload new data to "The Collector." Greg marvels at how technology has sped up data processing, also engaging college students and citizen scientists. He reminds us that the Golden Spike was just elevated to a National Historical Park: a new designation that will expand the site. The Bureau of Land Management is also expanding "without more employees," he adds, "not in an administration that is trying to shrink us." As a land manager, the BLM traditionally has raised national revenue through cattle

. . . rejuvenate, relational, renewable, repair, reparation, reservation, resilience, retreat, re-view, . . .

grazing, oil and gas leases. Rollbacks of conservation protections are increasing this activity alongside eco-recreation that, in many places, also jeopardizes Native American lands. "When future hikers come to viewsheds, are they going to see a grand vista or pump jacks?" he asks. Varied considerations have arisen at *Spiral Jetty* as partners have collaboratively curated an artwork on state lands. "One thing that I struggle with," Jaimi says, "is wanting to shout to the world how cool it is but not attract people to the tar seeps." She worries what might happen to the lake with increasing pollution and bioaccumulations of toxins in birds. "This is the heartbeat," she says of Great Salt Lake. "It's the place of transformation. The arteries that go into the lake look like what's going into your body," she says. "We have to honor what people were doing hundreds of years ago. Migrations here included the language of 'well-watered and eternal waters.' We still tell these narratives of ourselves—those narratives stick with us. But when there aren't eternal snows on mountains, it's hard for us to accept that change. 'We need to meter our water laws and not water down our laws,' as people say."

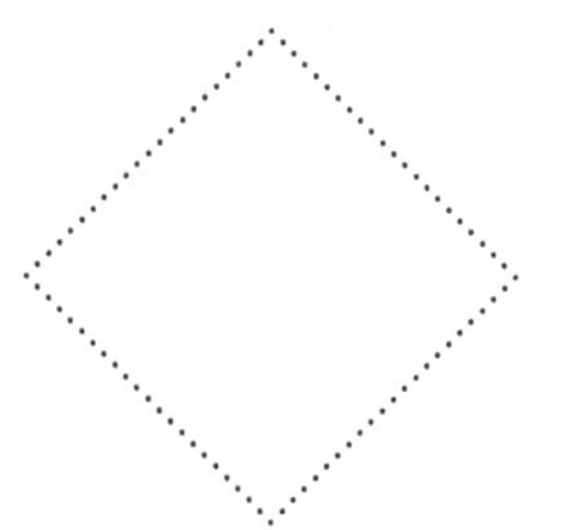

Rewind to 1896: *Now know you,* reads a land deed from the Utah Territory spelled out in cursive script: *To have and to hold the said tract of Land, with the appurtenances thereof unto the said X___ and to his heirs and assigns forever, subject to any vested and accrued water rights for mining, agricultural, manufacturing, or other purposes, and rights to ditches and reservoirs used in connection with such water rights.* The deed does not refer to the land's inhabitants prior to the 1862 Homestead Act, overwriting Indigenous lives, as it was written "To secure Homesteads to Actual Settlers on the Public Domain." Fast-forward to 1940: a typed letter from a lieutenant colonel in the Air Corps to the vice president of the Utah Audubon Society: "in view of the present National political situation which confronts our country, that the proper training of Air Corps personnel in bomb-

. . . sacred, sacrifice, salt lake, sanctuary, sediment sample, seep, seismic, shadow, species, Spiral Jetty, strata, stuck, surface, . . .

ing and aerial gunnery is of greater importance to the country as a whole than the preservation of certain bird life in this vicinity." Fast-forward to 2020: a co-authored commentary digitally published in *Nature Sustainability:* "As vignettes, the stories are unavoidably incomplete. Who does the imagining matters for which story is told, for what stories are enacted. The stories illustrate that biodiversity, climate and inequality are inseparable."

Like a gull getting stuck in a tar seep, we can get stuck in one way of perceiving the world. We may be lucky enough to avoid a stinky death trap—because it forces us to immediately react or else die—but we may not notice ourselves getting stuck in a mode of thought, in a single frame of reference, in a single story. This leads me to wonder about your very act of reading this book and even my writing it. Both practices indicate that we look for meaning in narratives. To rethink narratives, how can we rethink our place in the world? What happens when we change a singular story and its correlated metaphors to multiply possibilities and ask: What else are we *not* perceiving here and elsewhere? If we can't even listen well to human stories, what other stories are we not hearing?

When we return to Salt Lake City, the sun shines, after our day-long trek began in rain and snow. Honeycombed dirt coats the vans. We won't see each other again at Great Salt Lake but in cafés, offices, homes. I will soon head home to Washington, D.C., after a few last trips out West that will appear in my photo stream months later, mixed into a scrollable contact sheet, with a star-shaped tar seep, a gull flying overhead, wagon tracks in the sand, my footprints in tar. I will remember the broken glass of my iPhone: like dendritic seeps and cracks in dry earth that are not only signs of ending but continuation, expanding outside my frame of reference. I will wonder again how Great Salt Lake remembers, as a body of water accumulating to impact its existence over time. It cycles to snow and rain and runoff down the mountains back to the lake. Evaporation, condensation, precipitation, collection. The lake's story continues through seepage around tectonic plates, which quake and fracture, so raw oil creeps to the earth's surface as tar seeps: quiet reminders that we are all stuck together.

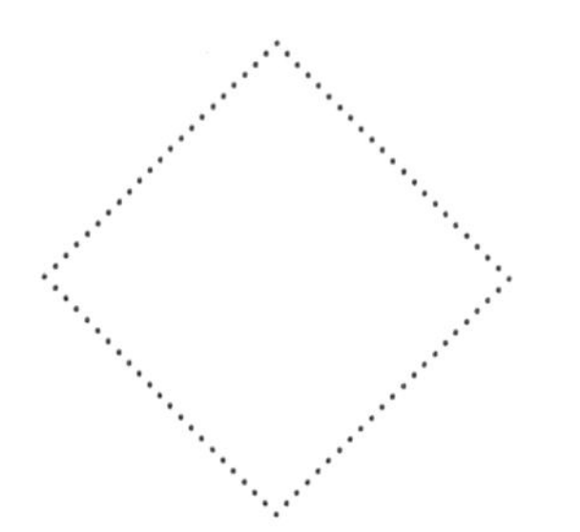

The ancient Maya greeting *in lak'ech—a la k'inn* has been translated as *I'm another you. You're another me.* "Both suggest a kind of communal ethos baked into how two strangers might regard each another," writes Hua Hsu in "The Search for New Words to Make Us Care about the Climate Crisis." "Naturally, there is a risk that borrowing someone else's language, terminology, or scraps of larger belief systems might feel like a fetishizing gesture . . . Perhaps, at a time of such stark extremes, there's something meaningful about language that describes transition, a state of in-betweenness . . . a story . . . with a dozen different endings, bound by a collective push to rethink what we resign to inevitability."

A few months after my last visit to the seeps, my five-year-old niece Riley and I travel with my parents to the Cascade Mountains in Washington state, where my great-uncle Fritz and great-aunt Gertrude used to live, where he used to watch ospreys nest and soar. I have visited the area since childhood, often seeing birds, but it is my first chance to visit with Riley, who lives in a suburb of San Francisco. On each visit I see the place anew. As I look through binoculars at the Wenatchee River, Riley starts to notice. Of her own accord, she starts a bird book, asking me to spell words that she can't yet read. In her letter-by-letter script, lines fill with their names. Over a few days her list grows: *OSPREY, TURKEY VULTURE, BALD EAGLE, CEDAR WAXWING, SONG SPARROW, RED-WINGED BLACKBIRD, ROBIN, GREAT BLUE HERON, MERGAN-*

. . . virtual, viscous, vision, vital, volatile, vulnerable, watershed, wild, wonder, world, worldview, . . .

SER, STELLAR'S JAY, and many more. Her excitement confounds me. Birds are not an interest that she shows at home. From the local public library, we check out a "nature backpack" filled with child-sized binoculars, a mini-compass, colored pencils, trail guides, and laminated books of animals, reptiles, and birds. By the end of the week, I awake to find her on the deck while the rest of the family sleeps, having pulled out a fold-up chair and blanket, swaddling herself in the chill to peer through "noculars" (as she calls them, having just learned the word), almost leaping out of her chair when she sees an osprey land on a neighboring tree. She points and repeatedly whispers its name in English. We wake my father, her paternal grandfather, to show him the bird. (Later, I will learn from her maternal grandmother that "osprey" in Korean is written as 물수리, pronounced "mulsoori," where "mul" (물) means "water.") Riley asks me to read aloud her bird list. Bird by bird, she compares each name with images in the laminated bird guide, as I play recordings of their calls from an app. She asks if we might see a *pelican:* a bird that she remembers seeing when she visited me months earlier in Salt Lake City. "Probably not here," I say. "But we'll keep looking and see what else we see."

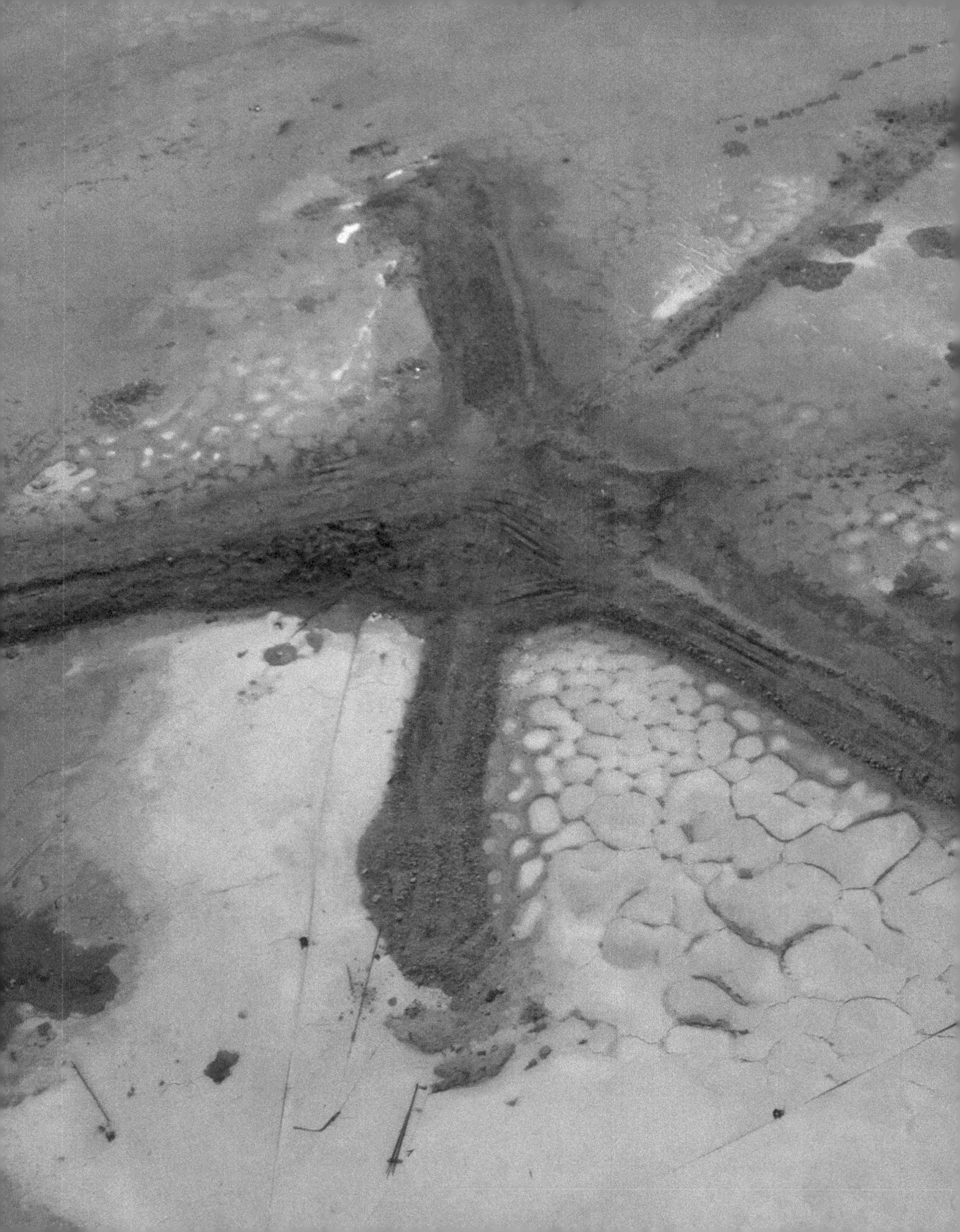

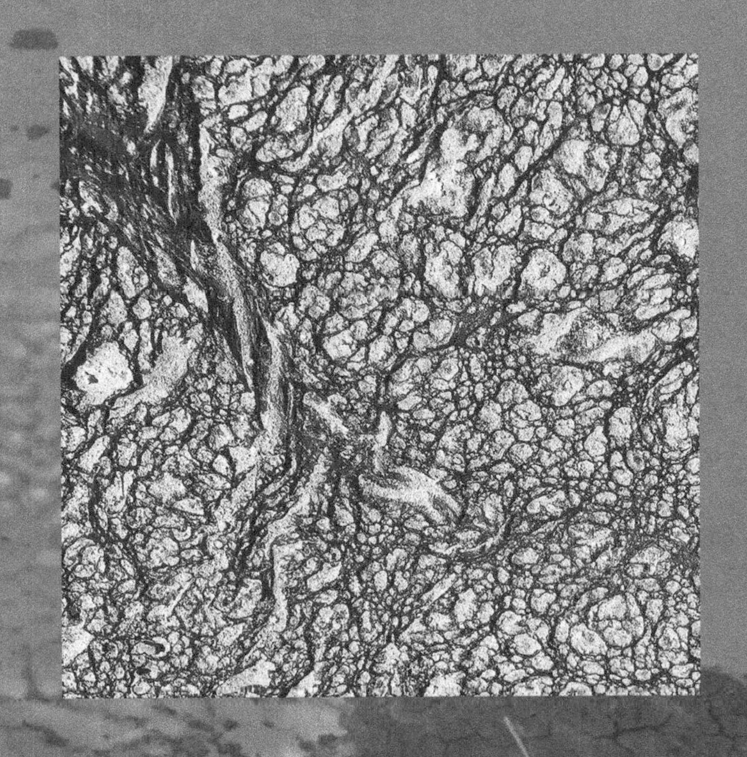

EARTH

POSTSCRIPT

What affirmative visions of the future
can the environmentalist movement offer,
visions that are neither returns to an imagined pastoral past
nor nightmares of future devastation
meant to serve as "cautionary tales"?
—Ursula Heise, *Imagining Extinction*

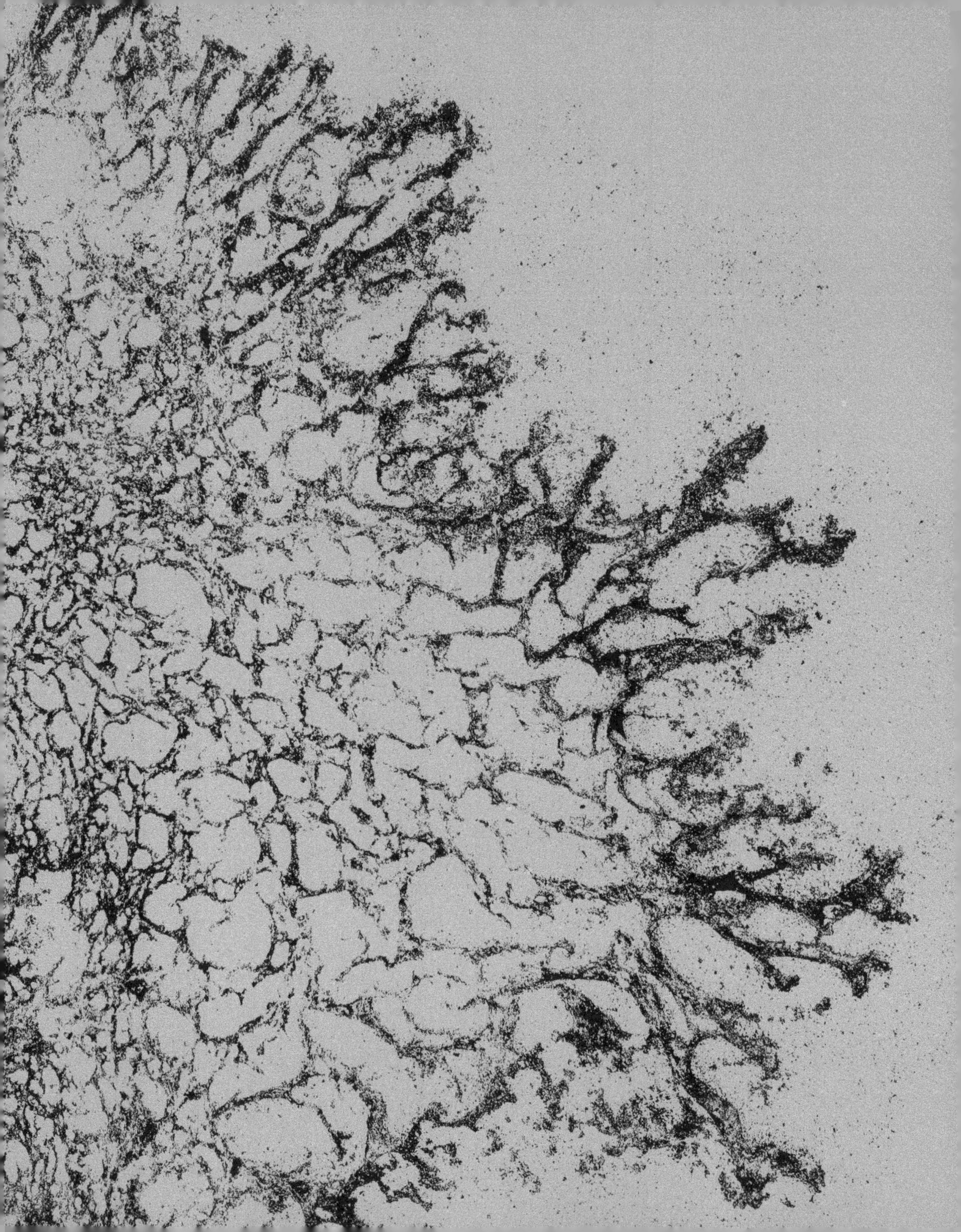

The Big Here

Once upon a time carried us away way back when: *once upon a time,* when the voice of a storyteller spun a spell that kept us listening with bated breath for what came next. Part of the lure came from listening to the voice and its pace, unspooling the larger story that seemed to emerge from thin air as a filament of silk. The silk started to web. Breath upon breath, line upon line, the voice conjured landscapes vast as kingdoms or palm-sized wonders as hummingbirds, winging into blur. Storms swept in. Clouds swallowed stars. We saw light aglow in windows, heard distant thunder, smelled sweet decay of hay and oncoming rain, opened our mouths to taste falling water, droplets covering flesh and melting as shedding skins, as we ran for shelter to a flood of warmth, and light, and . . .

Then.

Years sped ahead.

We heard different stories and more ways of telling, lived our lives not as linear paths but diverging junctures. Alternate lives arose through choice and chance.

Our lives shifted shapes of stories, even as we tried to fit things neatly into *Once upon a time*. By listening, more ways of telling arose: ways of living in this world. Diverse forms evolved through biodiverse contents. Senses veered toward lyric poetics and multilingual tongues, connecting ways of knowing beyond our own minds and hearts. Listening offered a quest of questions. At the interstices of these attentions were links so obvious as to be often overlooked, basic as breath.

Words carry breaths of a body into a book. Sentences set a pace. A comma invites a quick intake of air, where a semicolon is more of a gasp; a period outright pauses. The break of a paragraph inhales and exhales, sharing a pulse. Depending on what kind of story is being told, the rhythm casts a spell, pulling the mind or quickening the heart. A spell rests in spelling, as the alchemy of the alphabet, where letter by letter a word conjures a concept that we recognize in the world. The projected story reanimates what we see, hear, feel, taste, touch. As breaths bridge letters into words into sentences, between lines: absences become presences that press to the edges of our knowledges and perceptions. To write or read inhales and exhales, not in isolation but in relation, charged by the air that we breathe.

In our digitally driven age, the multisensory dimensions of writing sometimes press to the edges of the visual, to the point that we hold our breath. We sometimes forget "the eyes of the skin," to borrow a phrase of Finnish architect Juhani Pallasmaa, where all parts of us breathe—which becomes the basis of what we perceive and don't, of what we make and don't. Deficits of "focused vision" push us out of space into the role of disembodied spectators, while "peripheral vision" integrates us into space, aware of our environment and each other. Thinking back to paintings in caves, ancient handprints and ochre bison, images are not of sights alone: warming with fire, dirt on skin, reek of breath, taste of song, and shadows where senses blur. Perhaps in a cyclical sweep, genres may be expanding in three dimensions: to put the logic of the mind back inside the body, in the breath and beat of our hearts, in the hum and spin of the Earth.

As we grow disembodied in the digital age amid climate change, I have been weighing "essays spread out in space" and intermedia genres as ways to sound the gaps—not leading us into virtual realms but back into the world that we cohabit. As we injure Earth, we forget reciprocity and, in turn, injure each other. How can we cultivate the art of attention to better understand our actions and interactions, passive and active engagements of care and repair? How can we imagine the ripple effects of our smallest moves to be generous and rejuvenating? Can we "desegregate wilderness," as Jourdan Imani Keith urges, to integrate and bridge divisions? Or reseed collective dreams by "unprecedented listening"—as many voices echo—to interconnect stories organically as the plural roots of an aspen? What might emerge through reimagining our relationship with carbon to produce "buildings like trees" and "cities like forests"? Or renewable ways to listen? Trees are talking; birds are winging; whales are singing; stones are ringing. As our species has radically unmade the world, can we radically remake ourselves to allow each other and the Earth to heal?

This little book falls short of these aims; yet, its betweenness invites reading strategies beyond its pages in print and online, seeping into the wider world. The story itself is stuck as a kind of seepage: aiming toward fieldwork from afar, questioning inscription as it is written, this "essay" oscillates on the line of its declassification. Or, maybe this process is inherently characteristic of a genre of assaying, which reaches in and out of itself (and myself and yourself, where we meet ourselves as a species) to test our edges as we come into being:

Breathe.

Coda

Light.

Moments dissolve in the night.

Light.

Flashing in the dark: sparks against silhouetted trees.

Light, bright, light: in rapid succession, countless quiet flares.

The park engulfs us in a grove, on a grassy slope. Stars blink in and out, bright and fading, bright again, flitting as we move into darkness.

Trees fade; our eyes adjust.

Lights wick as candles on branches. Flickering, blinking: brighter, faster. Flashes quicken into rapid, fleeting constellations. Sparks pulse as an electric circuit, faster than neural charges, as if beyond the speed of light—not isolated but connected—surging through the thick, summer dark of the park.

Lightning bugs flash too fast for me to count. It is firefly season. Night. Above the trees, the sky spatters muted stars. A full moon glows.

I have never seen anything like this natural display of light: quiet wonder of the world, in an urban park in Washington, D.C., not wider than a city block. My husband and I are standing among woods near our home, away from downtown lights. This city is fringed with parkways, meandering slivers of green, around creeks that feed the Potomac River, the Chesapeake Bay, the Atlantic Ocean. Summer leafs and sprouts with every shade of green, tropical in its humid density.

For a few weeks each year, fireflies come to windows, to yards, to this park, slivered woodlands, rock-littered streams, and marshy meadows. Over the years, we have witnessed fireflies at dusk but had to learn to go deeper, into a grove of trees, to see them cluster and multiply beyond count. I soon realized their subtle light doesn't register in my camera, that I need to put that lens away, to be present. It took a few years of living here to discover how fireflies swarm—living light, bright in the dark, wonders in plain sight—right before us, here and now.

This ephemeral sight will stick in my memory longer than any photograph. Against this backdrop, I have written dozens of postscripts that I have deleted:

> about Nancy Holt's *Dark Star Park* (1984), which I bypassed on daily walks upon first moving to the D.C. area and never noticed, going cross-country to visit the *Sun Tunnels*, only recently witnessing *Dark Star Park*'s annual alignment of sun and spheres (nicknamed, to Holt's chagrin, "Arlington's Stonehenge");

> about other Land Art that we visit nearby, including *Always Becoming* by the New Mexico–based artist Nora Naranjo-Morse (Santa Clara Pueblo), that was installed off the National Mall in 2007: as a family transforming from the earth "at the cross section of asphalt and / fertile ground," in a region of ancestral lands of the Piscataway, Pamunkey, the Nentego (Nanichoke), Mattaponi, Chickahominy, Monacan, and the Powhatan cultures;

> about a family of four barred owls that encountered me one dawn in nearby woods (lands that were a former Civil War military battery, now national park), less than ten feet from my trail, circling me on branches, as if teaching bypassing humans how to listen to their juvenile calls;

about Green Bank Observatory in the 13,000-acre National Radio Quiet Zone, founded in 1956 in the shadow of the Cold War, less than a day's drive to West Virginia, where we walked around the radio observatory that first heard the center of our spiral-shaped galaxy of the Milky Way;

about a palm-sized baby panda, born in the National Zoo during a global pandemic, watched remotely by Panda Cam and claimed a "miracle" among endangered and vulnerable charismatic species, entangled with our future, as microbial lives swim underappreciated inside us and in saline seas;

about more and more and more:

But for a moment, here, in the dark of this urban park, we see thousands of quiet flashes: not in the past or future, but here and now. For this brief moment, we are not worried—about all that has been and will be, not diminished by fear or sorrow over the harrowing inequity and ecological destruction of the world, not distracted by the decomposing state of our bodies, which is the wonder that makes us human—in spite of or because of it, we are here: where light is possible, where we open our hearts and ears and can move deeper into the darkness and uncertainty and find light, beckoning us to move forward.

How big is here?

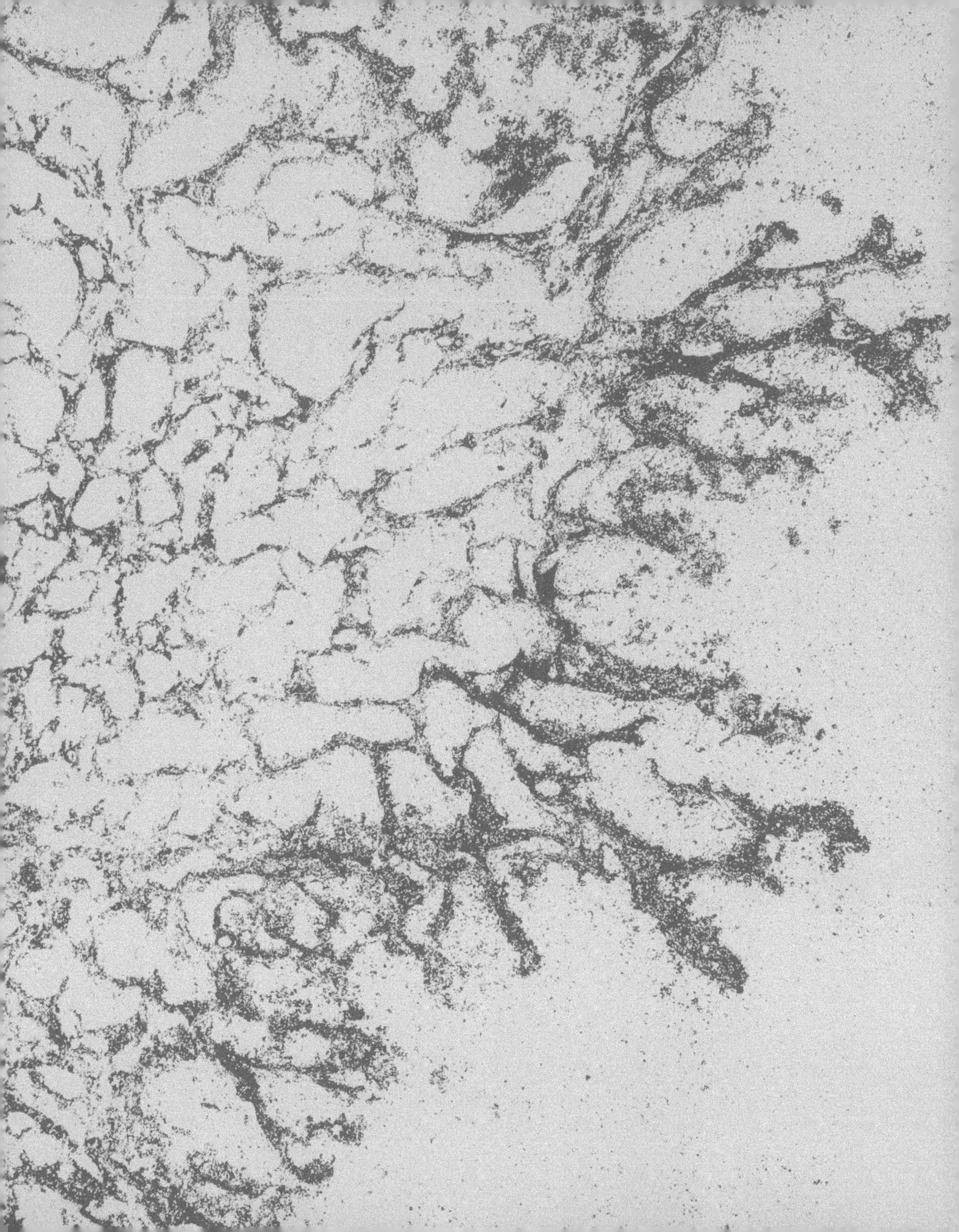

Field Notes

EXCAVATIONS

To be lost is to be fully present, and to be fully present is to be capable of being in uncertainty . . . which is totally unknown to you is usually what you need to find, and finding it is a matter of getting lost. — Rebecca Solnit, *A Field Guide to Getting Lost*

Watersheds gather, store, channel, schedule, disperse, concentrate, dilute, and . . . transform.

—Peter Warshall, from *Writing on Water*

Field Notes on Water
runoff from your watershed

1. Locate yourself in relation to a body of water. By what name(s) do you call this water?
2. Trace the water that you drink to its source.
3. Can you draw the boundaries of your watershed?
4. How many feet above sea level are you?
5. Where is the nearest ocean, and when is high tide?
6. When will be the next full moon?
7. What time are sunrise and sunset today? Where does the sun meet your horizon?
8. Find the North Star by night—or, by day, the direction of north.
9. From what direction do storms tend to come?
10. Where does rainwater from your roof go?
11. What was the amount of annual rainfall last year?
12. What forms of water (e.g., rain, snow, ice) do you encounter?
13. Where does gray water versus black water go from your home?
14. What is the composition of soil under your feet?
15. How far would you need to dig to reach water?

16. What spring wildflower is among the first to bloom here?
17. Name three native edible plants in your vicinity. What are their growing seasons?
18. Name two "invasive" species in your vicinity. When and how did they arrive?
19. Name four birds in your area. Are they migratory, or do they stay year-round?
20. When did you or your family migrate to this place? From where did your ancestors come?
21. How has your history and that of your family been shaped by water?
22. Does your memory of this place precede your lifetime, and how?
23. Who were previous inhabitants in this region, and how did they sustain themselves?
24. How many people currently live in your watershed?
25. How does water in your region assert force of life (for instance, by extremes, when was your last water event: a flood or a drought, Day Zero or . . .)?
26. Where is the nearest earthquake fault, and when did it last shift?
27. What geological processes shaped the landscape around you?
28. What geological or other natural features of your watershed have been considered sacred, now or in the past?
29. Beyond survival, what significance does water carry for you (culturally, spiritually, economically, politically, or otherwise)?
30. How are natural resources, including water, economically valued or extracted where you live?
31. In what ways is water represented around you? How does water flow—converge or diverge—through science, literature, art, and other bodies of knowledge?
32. How do you consider the shape and substance of your body related to bodies of water (e.g., aquifer, stream, river, lake, ocean)?
33. How much of your body consists of water? What percentage of the body of Earth is water?
34. From where is your home's energy, electricity or otherwise, generated?
35. When was the last time fire burned through the nearest wilderness, and where was that?
36. What sources of pollution inhabit the air that you breathe?
37. What species used to live here that have gone extinct, and how long ago did they die?
38. What former bodies of water used to exist in your region, but no longer do? Did they leave an imprint?

39. In recent years, did the answer(s) to any of these question(s) change, why and how?
40. Did any previous question(s) elicit hope or fear, or some other emotion, or no emotion at all? How do your emotions relate to the reality of others?
41. In what ways do you interact with water? How much water do you use daily?
42. How is your life enhanced by water? In what ways does water bring you joy or pleasure?
43. Do you suffer from any issue related to water, quality or scarcity, or something else?
44. Who do you know (a loved one, a stranger, another species) who suffers because of an issue related to water? How does suffering shape you, directly or indirectly?
45. In your region, what projections about water have been made for ten, twenty, and fifty years?
46. In the past, what plant forms dominated your surrounding landscape, 100 years ago and 10,000 years ago?
47. What places on other continents share similar characteristics (e.g., temperature, rainfall, vegetation) to where you live?
48. If you were to share a story about a body of water about which you care, what body of water would you choose? How would you tell their story, in words or by other means?
49. How do you listen to water?
50. What are three small acts that you can do, here and now, to care for water? How do these small acts reflect your hopes for the future of water?
51. What other questions might you add here?
52. How will you live today?

Please add more questions.

(a variation and expansion of "The Big Here Quiz"—see *Field Notes for Further Reading*)

Photographers have struggled with the beauty of ugliness . . . Wounds on the land, sculpturally destroyed or picturesque—graphic aerial shots or colorfully polluted waters—can be so striking that the message is overwhelmed and misery or horror is merely estheticized . . . On the other hand, beauty can powerfully convey difficult ideas by engaging people when they might otherwise turn away.

—Lucy Lippard, *Undermining*

Field Notes on Looking
a visual itinerary

All photographs are by the author unless otherwise noted.

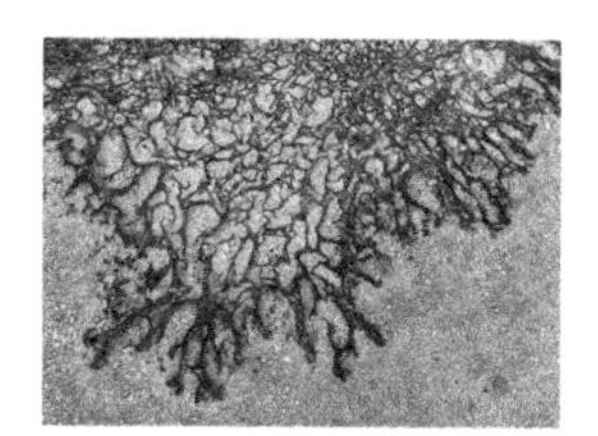

"Field Note: frozen melting seep lacework," Rozel Point, Great Salt Lake, Utah, 02/12/2018.

"Field Note: Pelican walking on tar seep with bones," Rozel Point, Great Salt Lake, Utah, 08/24/2018, camera trap image courtesy of Great Salt Lake Institute, Westminster College.

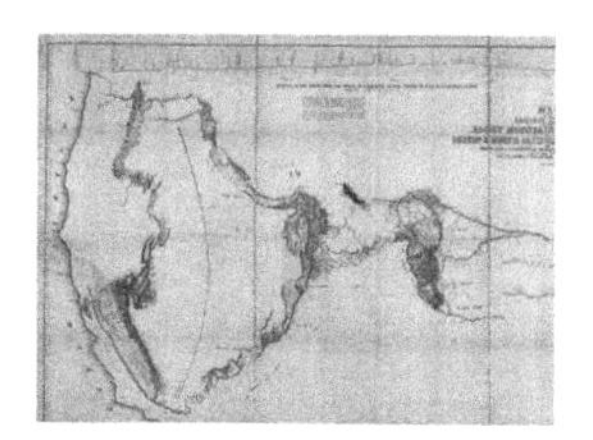

John Charles Frémont, et al., *Map of an exploring expedition to the Rocky Mountains in the year 1842 and to Oregon & north California in the years 1843–44* (Washington, D.C.: U.S. Senate, 1844). Map. Library of Congress: 96688042.

"Field Note: frozen melt of tar seep," Rozel Point, Great Salt Lake, Utah, 02/12/2018.

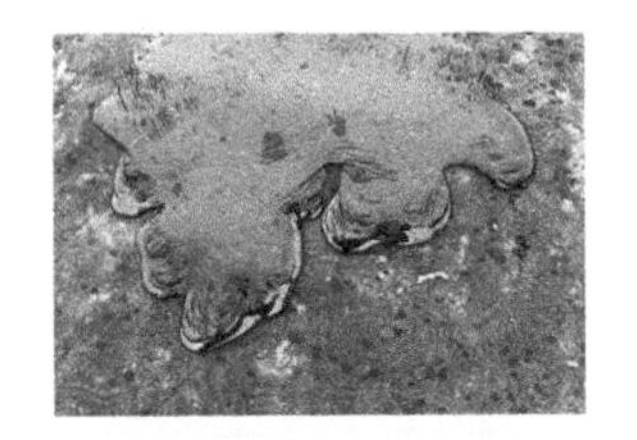

"Field Note: Pelican death assemblage with tags," Rozel Point, Great Salt Lake, Utah, 02/12/2018.

"Field Note: Live pelicans gathering around death assemblage in tar seep," Rozel Point, Great Salt Lake, Utah, 08/13/2018, camera trap image courtesy of Great Salt Lake Institute, Westminster College.

"Field Note: Gypsum from Rozel Point, Utah," Washington, D.C., 11/2020.

"Field Note: Great Salt Lake, evaporating," Rozel Point, Great Salt Lake, Utah, 04/27/2018.

"Field Note: star-shaped seep with barn owls," Rozel Point, Great Salt Lake, Utah, 02/12/2018.

"Field Note: circling north star-shaped seep," Rozel Point, Great Salt Lake, Utah, 02/12/2018.

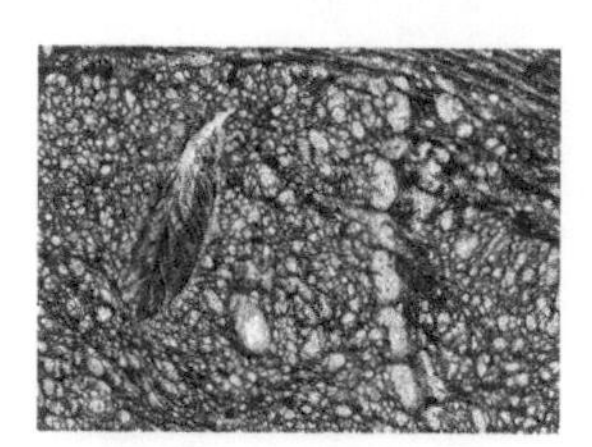

"Field Note: barn owl feather constellated with bone," Rozel Point, Great Salt Lake, Utah, 02/12/2018.

"Field Note: barn owl feathers," Rozel Point, Great Salt Lake, Utah, 02/12/2018.

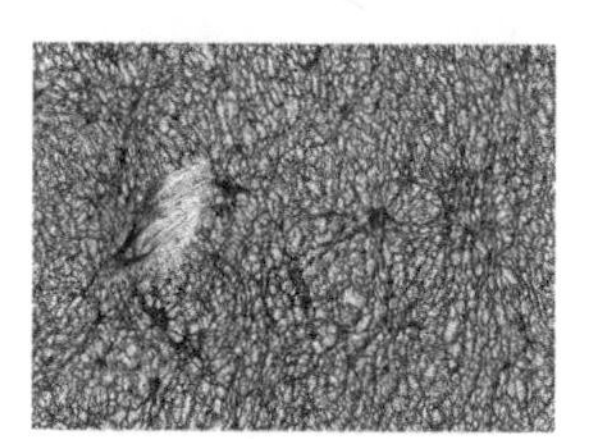

"Field Note: barn owl feathering bone," Rozel Point, Great Salt Lake, Utah, 02/12/2018.

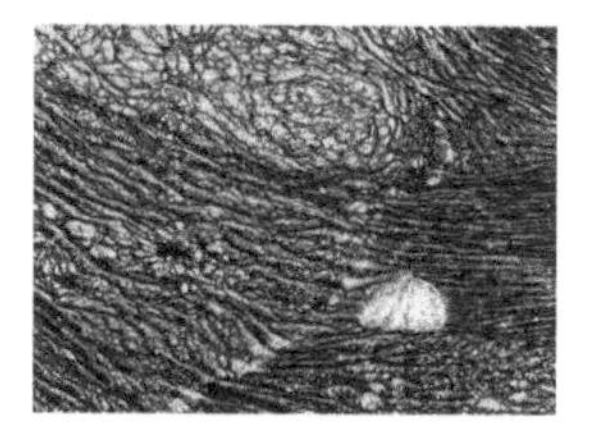

"Field Note: barn owl feather fragment," Rozel Point, Great Salt Lake, Utah, 02/12/2018.

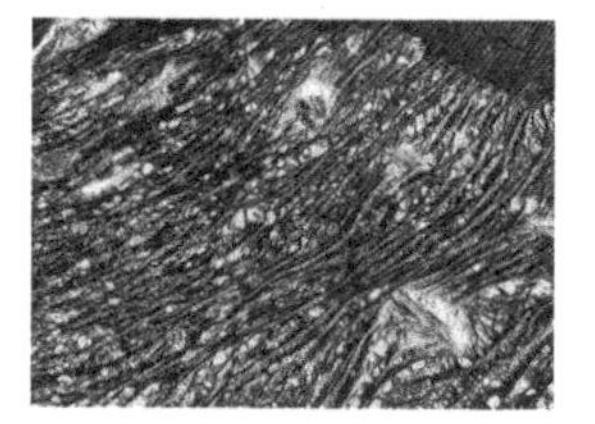

"Field Note: barn owl fragments," Rozel Point, Great Salt Lake, Utah, 02/12/2018.

"Field Note: close-up of star-shaped seep," Rozel Point, Great Salt Lake, Utah, 02/12/2018.

"Field Note: circling star-shaped seep," Rozel Point, Great Salt Lake, Utah, 02/12/2018.

"Field Note: star-shaped seep on mudflat," Rozel Point, Great Salt Lake, Utah, 02/12/2018.

"Field Note: view from bluff above *Spiral Jetty*," Rozel Point, Great Salt Lake, Utah, 02/12/2018. Robert Smithson, *Spiral Jetty* (1970); © Holt/Smithson Foundation and Dia Art Foundation / VAGA at Artists Rights Society (ARS), NY. Photo: G.E.H.

Great Salt Lake, Utah, satellite view. Map data: Google, TerraMetrics, 2020.

Great Salt Lake as seen from International Space Station, photograph by Alexander Gerst, European Space Agency, public domain, 2014.

"Field Note: aerial arrival (over Great Salt Lake mineral ponds, surrounds of Salt Lake City, and Wasatch Mountains)," Great Salt Lake, Utah, 12/29/2017.

"Field Note: *Spiral Jetty* above Great Salt Lake waterline," 10/15/2017. Robert Smithson, *Spiral Jetty* (1970); © Holt/Smithson Foundation and Dia Art Foundation / VAGA at Artists Rights Society (ARS), NY. Photo: G.E.H.

"Field Note: *Spiral Jetty* with Great Salt Lake water," 03/24/2019. Robert Smithson, *Spiral Jetty* (1970); © Holt/Smithson Foundation and Dia Art Foundation / VAGA at Artists Rights Society (ARS), NY. Photo: G.E.H.

"Field Note: aerial flyover (over Great Salt Lake mineral ponds through layover in Salt Lake City from Sacramento, CA, to Washington, D.C., after grandmother's death in Grass Valley, CA)," Great Salt Lake, Utah, 07/26/2015.

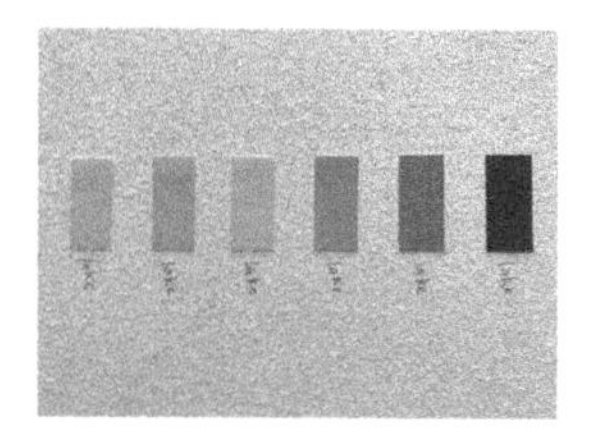

"Field Note: lake, lake, lake, lake, lake, lake," 05/03/2018, from Spencer Finch, American, born 1962, *Great Salt Lake and Vicinity* (detail), 2017, 1,132 ready-made Pantone chips and pencil, commissioned by the Utah Museum of Fine Arts, University of Utah, Salt Lake City, purchased with funds from The Phyllis Cannon Wattis Endowment Fund, UMFA2018.4.1. © Spencer Finch. Photo: G.E.H.

"Field Note: Pair of pelicans crossing tar seep," Rozel Point, Great Salt Lake, Utah, 08/14/18, camera trap image courtesy of Great Salt Lake Institute, Westminster College.

"Field Note: *Sun Tunnels,* Great Basin Desert, Utah, 04/14/2018. Nancy Holt, *Sun Tunnels* (1973–76); © Holt/Smithson Foundation and Dia Art Foundation / VAGA at Artists Rights Society (ARS), NY. Photo: G.E.H.

"Field Note: Rainbow Bridge National Monument, Utah," 04/17/2019.

"Field Note: sign at Lucin on migrating birds, with bullet holes," Great Salt Lake, Utah, 04/14/2018.

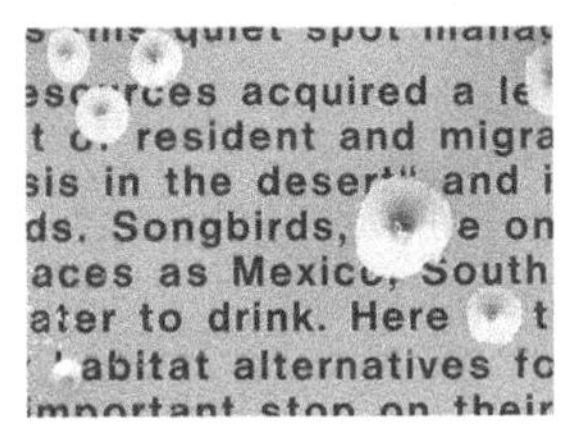

"Field Note: aerial view of Rio Tinto Kennecott Bingham Canyon Mine," Utah, 04/07/2019.

"Field Note: aerial view of Kennecott Copper Smelter smokestack," Great Salt Lake, Utah, 01/04/2019.

"Milky Way above Sierra Nevada," Nevada City, California, 08/11/2020. Photo by author's father, Martin F. Ernster.

"Field note: rusting equipment near the tar seeps," at Rozel Point, Great Salt Lake, Utah, 04/14/2018.

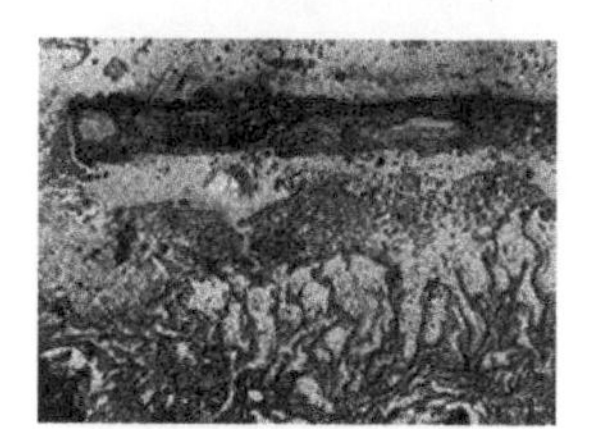

"Field note: abandoned attempts at oil drilling," Rozel Point, Great Salt Lake, Utah, 04/14/2018.

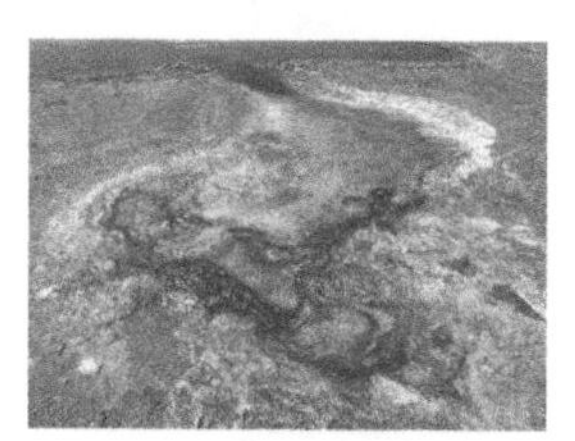

"Field Note: shoreline with halophiles by *Spiral Jetty,*" Rozel Point, Great Salt Lake, Utah, 04/27/2018.

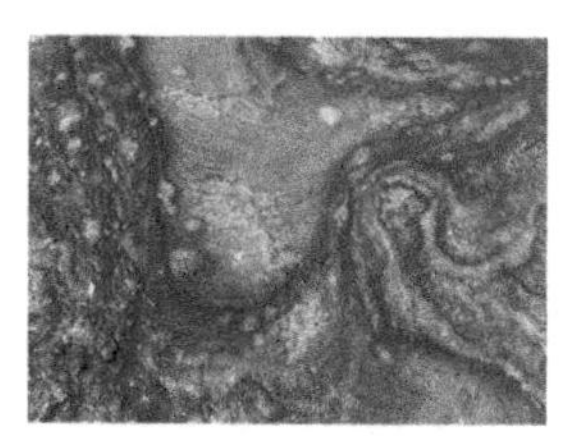

"Field Note: close-up halophiles in saltwater by *Spiral Jetty,*" Rozel Point, Great Salt Lake, Utah, 04/27/2018.

"Field Note: visitors to *Spiral Jetty,*" Rozel Point, Great Salt Lake, Utah, 04/27/2018.

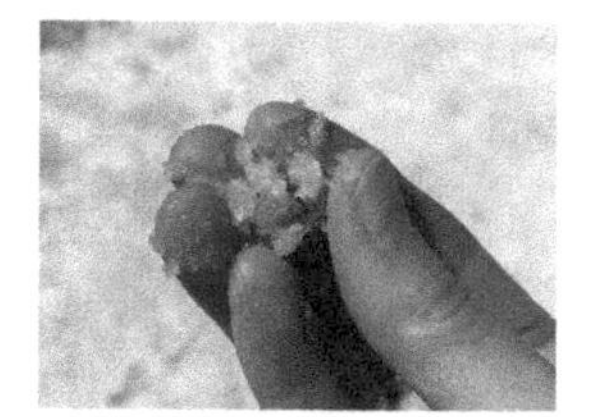

"Field Note: touching salt from Great Salt Lake," Rozel Point, Great Salt Lake, Utah, 04/27/2018.

“Field Note: Kara Kornhauser holding camera trap,” Rozel Point, Great Salt Lake, Utah, 04/27/2018.

“Field Note: gull flying over ‘pelican death assemblage,’” Rozel Point, Great Salt Lake, Utah, 04/27/2018.

“Field Note: wagon tracks from group trek to set camera traps at tar seeps,” Rozel Point, Great Salt Lake, Utah, 04/27/2018.

“Field Note: view through rotting pylons,” Rozel Point, Great Salt Lake, Utah, 04/27/2018.

“Field Note: return to tar seep with barn owls, star melted beyond recognition,” Rozel Point, Great Salt Lake, Utah, 04/27/2018.

“Field Note: once feather, now bone,” Rozel Point, Great Salt Lake, Utah, 04/27/2018.

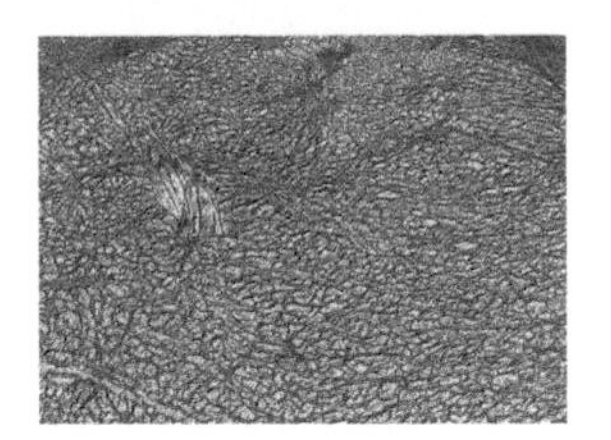

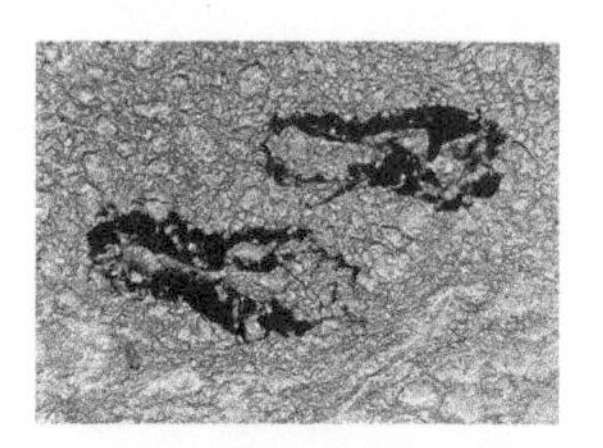

"Field Note: my footsteps in once-star-shaped seep," Rozel Point, Great Salt Lake, Utah, 04/27/2018.

"Field Note: disappearing footsteps in once-star-shaped seep," Rozel Point, Great Salt Lake, Utah, 04/27/2018.

"Field Note: leaving the tar seeps by foot," Rozel Point, Great Salt Lake, Utah, 04/27/2018.

"Field Note: trio of gulls flying over 'pelican death assemblage' and landing past the edge on mudflat," Rozel Point, Great Salt Lake, Utah, 04/27/2018.

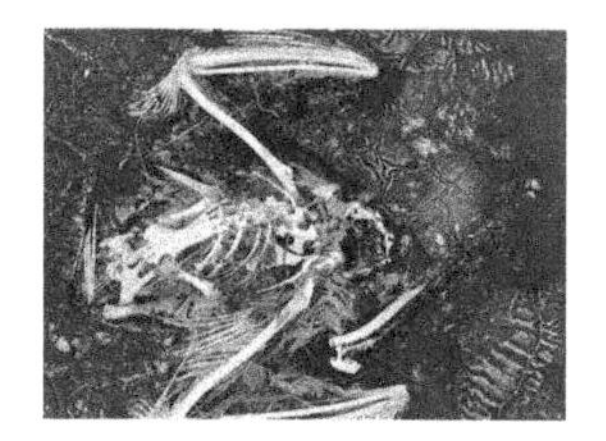

"Field Note: Pelican skeleton with splayed wings stuck in tar seep," Rozel Point, Great Salt Lake, Utah, 03/24/2019.

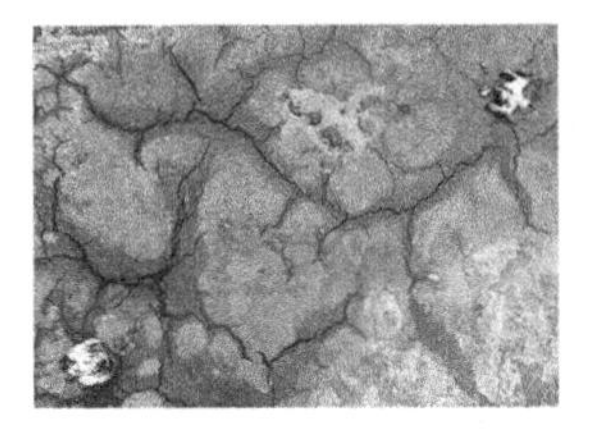
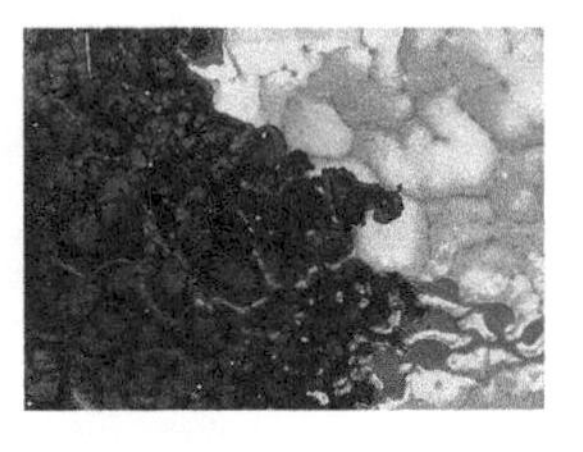
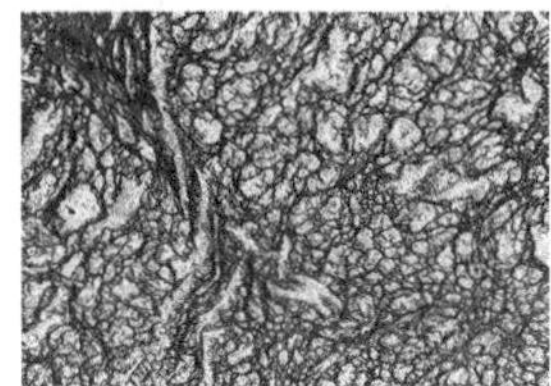

"Fossils in the Making" intermix drone photographs by the Utah Geological Survey (UGS), with thanks to Christian Hardwick, along with drone and other photographs by G.E.H.

"Field Note: Aerial view of pylons at tar seeps," Rozel Point, Great Salt Lake, Utah, 06/12/2021, drone photograph by G.E.H.

"Field Note: Aerial view of tar seeps, abandoned oil jetty, and mudflats," Rozel Point, Great Salt Lake, Utah, 06/12/2021, drone photograph by G.E.H.

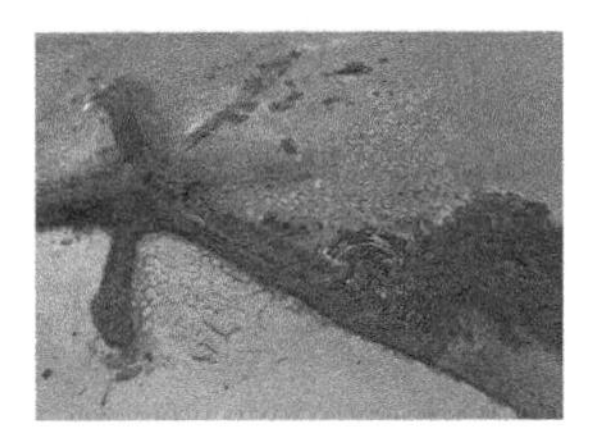

Additional photographic and textual Field Notes are available as digital seepage accessible by QR codes printed in this book (including the following links by the author):

https://www.gretchenhenderson.com/tarseeps

https://becoming.ink/pelican/

https://holtsmithsonfoundation.org/here-where-beyond-dark-star-park

https://luchoffmanninstitute.org/reimagining-biodiversity-narratives-and-pandemics/

https://www.nature.com/articles/s41893-020-0587-5

https://exhibits.lib.utah.edu/s/mining-the-west/page/toward-birds-eye

More QR codes are available on a DIY downloadable and printable sheet of stickers at https://www.gretchenhenderson.com/life-in-the-tar-seeps

Since QR codes in *Life in the Tar Seeps* are created with the intention of a minimal digital footprint, a reader is unlikely to get stuck and is encouraged to tune away from screens toward the world wherever they (we) are.

Field Notes on Method
fossils in the making

The printed book of *Life in the Tar Seeps* offers a field guide to get lost in the surreal, vulnerable, natural and manmade landscape of the tar seeps beside Robert Smithson's *Spiral Jetty* at Great Salt Lake, Utah, spiraling through North America across the Earth.

The book's print and digital integrations mimic the tar seeps, as the story melts into fragments around fossils-in-the-making. Depending on how a visitor accesses the site of Rozel Point in these pages, the tar seeps can be encountered across different media: as a book or virtual gallery, in print or online. As the narrative progresses, the printed text intermixes with QR codes: "quick response" grids that launch supplemental representations and resources when activated through a mobile device.

To figuratively excavate *Life in the Tar Seeps,* scan the grids with a QR-code reader on a mobile device to be directed to virtual paths toward and away from the site.

In years to come, after the QR codes in this book are no longer readable by mobile devices, they will become "fossils" of technology. As the book acts as a verbal and visual field guide, it also offers a tactile invitation. By representing some extended ephemera (text, photographs, eco-cinema) as QR codes, the book gestures toward its inevitable (de)composition while providing a small set of choices for navigating: to act or to retreat beyond these pages. Handling a book and a mobile device simultaneously, the very process of reading is slowed: as if weighing the worth of these choices.

Additionally, blank boxes are included to indicate further directions—now and in the future—whether a reader wishes to incorporate downloadable stickers of supplemental QR codes or make their own, to expand the narrative beyond these pages and stick together other overlooked places that might help us care about our interconnected planetary home.

Digitally, the presence of technology "fossils" alludes to a quality of "stickiness": a characteristic reminiscent of tar seeps. Digital stickiness is a sought quality in web design, gaming, and interactive virtual projects. A "sticky" web site compels visitors to linger and return often. Tar seeps are likewise sticky, melting as "death traps," where a step can fatally trap an unsuspecting wayfarer. Despite obvious limits of this comparison, a question emerges: to what extent does virtual stickiness lead our species into a kind of death trap?

When I first used the technology of QR codes in a novel (published in 2011) while experimenting with what I called "melting" books (from frozen to liquid states, decomposing and deforming across genres and media), QR codes were little used, so most readers were unfamiliar with the black-and-white grids. Now, a decade later, QR codes are ubiquitous and even passé: what I like to think of as "fossils-in-the-making." The ones in this book are used simply akin to marginalia, annotations, or side commentary. Static QR codes are free to create, yet they expire over time. Accessible at one moment, they later become unreadable and, in essence, become frozen (another resemblance to tar seeps, in winter).

Even if a reader does not have a mobile device, they will not be shortchanged bypassing the QR codes/augmented experience (i.e., not getting stuck) in *Life in the Tar Seeps.* The presence of these grids in the final chapter is not meant to divert the narrative but to signal there are options for a "quick response": to extend the printed book, virtually transporting a viewer to various fields akin to digital tributaries or runoff from Great Salt Lake. The paths of these QR codes are deliberately minimal: to avoid stickiness. There is also white space for readers to add ecological stories (broadly defined) of which they are part that deserve attention. In a small way, I hope this growing gesture invites readers not only to follow stories spiraling around the tar seeps at Great Salt Lake, but also to put down the book and reconsider other quick and slow responses to the climate crisis across the world, wherever we are.

Fossilizing footnote: Against incontrovertible proof of climate change, narratives have largely fossilized around apocalypse, prophecy, elegy, and tug-of-warring tropes of progress and loss. Shapes of stories recur to mark the edges of our fears, so our tellings fall into predictable patterns and separate us from the animals that we are. Humans are slow to see our contributions to ecological destruction that can't be neatly predicted, veering increasingly from past patterns into new uncertainties. Visually driven literacy in the digital age makes us less dependent on multisensory, embodied knowledges that help us to interact with the world beyond words. Data tries to convince our brains yet paradoxically numbs emotions. Even as we face abounding facts, collective denial has grown to frame humans as passive victims rather than active agents. Most stories arise from fear of death rather than awe of life—its aliveness, volatility, mortality—more aptly navigated through Indigenous tribalographies, Aboriginal songlines, trickster narratives, timeworn performative and speculative forms indivisibly interconnecting people with landscape. Rather than adapt narratives to accommodate the unexpected, many of us humans fall back on stories that want for control and reinforce tropes: the individual over communal, the human over nonhuman, and other reductive separations. In the process of trying to fix or fit into story forms, we grow less equipped to cope with metamorphosis, especially when natural forces raise their voices and howl. Amid climate events like hurricanes and forest fires and other forces that increasingly amplify dying, what other forms of narratives and poetics of space are possible that might reintegrate us as a species into living? What other questions grow between these lines?

Listen:

Field Notes on Listening
a reader's response

Can you feel your breath?

What knowledge does your body carry?

What sensations pull your awareness into, or away from, your body?

What parts of your body do you notice or take for granted?

How do you imagine other bodies around you might feel?

Can you hear more hearts beating?

How else do you feel and hear?

To read beyond this book,
close these pages and listen through all your senses
to the surrounding world.

Lost and Found

< INSERT >

< INSERT >

< INSERT >

< INSERT >

< INSERT >

< INSERT >

A word is as alive as a bird.
—Lyanda Lynn Haupt, *Rooted*

Language is fossil poetry.
—Ralph Waldo Emerson, "The Poet"

Mud, salt crystals, rocks, water.
—Robert Smithson, "The Spiral Jetty"

UnIndexing
seeping through the limits of classification

: abandoned, accident, aerial, aesthetics, agency, agricultural runoff, alive, American West, ***American Avocet, American Bittern, American Coot, American Crow, American Golden-Plover, American Goldfinch, American Kestrel, American Robin, American White Pelican, American Pipit, American Tree Sparrow, American Wigeon,*** anger, Anthropocene, archaeology, archive, arid, arterial, artwork, apocalyptic, articulation, ***Ash-throated Flycatcher,*** asphalt, *Asphalt Rundown,* **In the high desert of Utah at Great Salt Lake, a series of tar seeps spreads across the earth.** bacteria, ***Baird's Sandpiper, Bald Eagle, Bank Swallow, Barn Owl, Barn Swallow, Barrow's Goldeneye,*** basalt, basin, Bear River Massacre, Bear River Migratory Bird Refuge, ***Belted Kingfisher,*** Binagadi asphalt lake (Azerbaijan), biodiversity, biorhythm, birds, bitumen, black gold, ***Black Tern, Black-bellied Plover, Black-billed Magpie, Black-capped Chickadee, Black-chinned Humming-***

BIRD, BLACK-CROWNED NIGHT-HERON, BLACK-HEADED GROSBEAK, BLACK-NECKED STILT, BLUE-GRAY GNATCATCHER, BLUE-WINGED TEAL, BOBOLINK, BOHEMIAN WAXWING, BONAPARTE'S GULL, bones, brea, breathing, *BREWER'S BLACKBIRD, BREWER'S SPARROW,* brine shrimp, *BROAD-TAILED HUMMINGBIRD, BROWN-HEADED COWBIRD, BUFFLEHEAD, BULLOCK'S ORIOLE,* burial, burning water, *BURROWING OWL,* byproduct, **Nicknamed 'death traps,' also called *oil* or *petroleum seeps*, tar seeps act like sticky flypaper and creep up from tectonic fractures at seismic spots across the planet.** *CACKLING GOOSE, CALIFORNIA GULL,* camera trap, *CANADA GOOSE, CANVASBACK,* capped well, carbon dioxide, Carpinteria Tar Pits and Oil Seeps (CA), *CASPIAN TERN, CATTLE EGRET,* cattle guard, caulk, causeway, caves, *CEDAR WAXWING,* CLUI (Center for Land Use Interpretation), *CLARK'S GREBE,* classification, *CLIFF SWALLOW,* climate change, climate crisis, circumnavigate, collective, colonial-settlement, colonization, Colorado River Watershed, combustion, *COMMON GOLDENEYE, COMMON LOON, COMMON MERGANSER, COMMON MOORHEN, COMMON NIGHTHAWK, COMMON POORWILL, COMMON RAVEN, COMMON TERN, COMMON YELLOWTHROAT,* composition, conservation, constellation, *COOPER'S HAWK, CORDILLERAN FLYCATCHER,* crude oil, cultural history, **Over centuries, tar has accumulated fatal and extractive associations, often accompanied by oil drilling.** danger, *DARK-EYED JUNCO,* dark skies, death trap, decenter, decolonize, decomposition, deep time, derrick, desert, desertification, Dia Art Foundation, digital age, disarticulate, disaster, DNR–FFSL (Department of Natural Resources–Forestry, Fire and State Lands), documentation, *DOUBLE-CRESTED CORMORANT, DOWNY WOODPECKER,* downwinders, drains, drills, drips, drone, drought, dry playa, *DUNLIN, DUSKY FLYCATCHER,* dust, dystopia, **Culturally, tar also has stuck things together:** *EARED GREBE,* Earth, earthwork, *EASTERN KINGBIRD,* ecology, ecotone, elegy, element, elsewhere, embodied, empathy, energy, entropy, environment, *EURASIAN WIGEON, EUROPEAN STARLING,* evaporation ponds, *EVENING GROSBEAK,* evidence, evolution, excavation, extinction, extraterrestrial, **breaking down limited classifications of death** fallout, fault, fault line, feathers, *FERRUGINOUS HAWK,* field guide, fill-in-the-blank, Finch (Spencer), fissure, flammable, flow, flyway, focal point, footprints, *FORSTER'S TERN,* Fort Sill Tar Pits (OK), fossil, fossilize, fossil fuels, fragment, *FRANKLIN'S GULL,* frozen, futures, **to reveal underlying life, telling a tale of place-in-time:** *GADWALL,* gas, geology, geothermal, *GLAU-*

cous Gull, global, ***Golden Eagle,*** Golden Spike, Goshute, ***Grasshopper Sparrow,*** gravity, ***Gray Flycatcher,*** Great Basin, ***Great Blue Heron, Great Egret, Great Horned Owl,*** Great Salt Lake, GSLEP (Great Salt Lake Exploration Platform), GSLI (Great Salt Lake Institute), ***Greater Scaup, Greater Yellowlegs, Green-tailed Towhee,*** grief, ground, groundwater, gull, Gunnison Island, gypsum, **where an unsuspecting animal (like a pelican) following a trail over seasons (as tar freezes and melts) can get stuck.** habitat, halophiles, ***Hammond's Flycatcher,*** harvest, hawk, hazard, healing, heartbeat, heat, heliography, here, ***Hermit Thrush, Herring Gull,*** Holt (Nancy), ***Hooded Merganser,*** hope, horizon line, ***Horned Grebe, Horned Lark, House Finch, House Sparrow, House Wren,*** human, human activity, humility, hydrocarbon, **From tar pits in Los Angeles (with ancient mammoths, giant sloths, and saber-toothed cats), to tar seeps in the Talara Desert of Peru (with ancient songbirds), natural asphalt can fatally preserve.** I, igneous, incomplete, Indigenous, Industrial Revolution, injury, injustice, inscription, interdependent, international, intervention, inventory, inversion, IPCC (Intergovernmental Panel on Climate Change), islands, **In the reputedly 'dead sea' of Great Salt Lake, tar seeps lie near the convergence of two of the four major migratory bird flyways of North America** jeopardize, jetty, Jordan River, journalism, jumbled, juncture, juxtaposition, **and beside Robert Smithson's famous earthwork coiling into the seeming sea.** Kalahari Desert (Lake Makgadikgadi, Botswana), ***Killdeer,*** kin, knowing, knowledges, **Smithson chose the site for *Spiral Jetty* (1970) because of the tar seeps.** La Brea Tar Pits (CA), lake, Lake Bermudez (Venezuela), Lake Bonneville, lake levels, land art, Land Arts of the American West, landscape, ***Lapland Longspur, Lark Bunting, Lark Sparrow,*** larvae, ***Lazuli Bunting,*** LDS (Church of Latter Day Saints), leased lands, ***Least Sandpiper,*** "leave no trace," ***Lesser Goldfinch, Lesser Scaup, Lesser Yellowlegs,*** liquid, lives, local, ***Loggerhead Shrike, Long-billed Curlew, Long-billed Dowitcher, Long-eared Owl, Long-tailed Duck,*** lubricant, **Like a massive ink blot on the planet, tar seeps seem to defy language,** ***MacGillivray's Warbler, Mallard, Marbled Godwit, Marsh Wren,*** manufacture, Mars, McKittrick Tar Pits (CA), meander line, ***Merlin,*** metadata, metamorphic, metaphor, methane, microbialite, microenvironments, microseisms, migratory birds, military testing, Milky Way, mine, mineral, mineral pitch, mineral rights, mining, mortal, Mother

Earth, ***Mountain Bluebird, Mountain Chickadee, Mourning Dove,*** muck, mud, mudflat, mummify, myth, **as we grope to articulate qualities:** national, narrative, NASA (National Aeronautics and Space Administration), ***Nashville Warbler,*** natural gas, natural resources, nature, negation, Nevada Test Site, non-invasive, non-site, North America, ***Northern Flicker, Northern Goshawk, Northern Harrier, Northern Mockingbird, Northern Rough-winged Swallow, Northern Shrike,*** nuclear fallout, **falling back on the fossilized vocabulary of paleontology, to migratory biodiversity of ornithology, to the economic value of the oil.** object, objective, observations, ode, odor, oil, oil company, oil field, oil spill, oil well, ooze, ***Orange-crowned Warbler,*** ornithology, ***Osprey,*** overlooked, overwritten, owl, oxygen, **As I acquire terms,** ***Pacific Loon, Pectoral Sandpiper,*** paleontology, past, pastoral, pathway, paved, pelicans, perception, ***Peregrine Falcon,*** perspective, petroleum, photography, ***Pied-billed Grebe, Pine Siskin,*** pipe, pit, pitch, Pitch Lake (Trinidad and Tobago), playa, Pleistocene, ***Plumbeous Vireo,*** poetics of space, point of view, politics, porous, posthumous, ***Prairie Falcon,*** precipitation, prehistory, preservative, preserve, primal, project, projection, protect, **the terrain evades classification. A tar seep swallows all languages that attempt to describe it.** quakes, questions, QR-code (quick response), **Bacteria break it down.** reclamation, recover, ***Red Knot,*** Red List, ***Red-breasted Merganser, Red-breasted Nuthatch, Redhead, Red-tailed Hawk, Red-winged Blackbird,*** Refuge, refuge, regional, rejuvenate, relational, reminder, renewable, repair, reparation, reservation, resilience, resurgence, resurrection, retreat, re-view, ***Ring-billed Gull, Ring-necked Duck, Red-necked Phalarope, Ring-necked Pheasant,*** river runoff, rock, ***Rock Pigeon, Rock Wren, Ross's Goose, Rough-legged Hawk,*** Rozel Point, ***Ruby-crowned Kinglet, Ruddy Duck, Ruddy Turnstone, Rufous Hummingbird,*** **As some bacteria even eat tar, they suggest hope for cleaning up future oil spills, and something more philosophical:** ***Sabine's Gull,*** sacred, sacrifice, ***Sage Thrasher,*** sagebrush, saline, salt, salt flat, sand, ***Sanderling, Sandhill Crane, Savannah Sparrow, Say's Phoebe,*** science fiction, sediment sample, "sedimentation of the mind," seepage, seiche, seismic, ***Semipalmated Plover, Semipalmated Sandpiper,*** shadow, ***Sharp-shinned Hawk,*** shoreline, ***Short-billed Dowitcher, Short-eared Owl,*** Shoshone, sink, slow reading, smell, Smithson (Robert), ***Snow Bunting, Snow Goose, Snowy Egret,*** snowmelt, ***Snowy Plover,*** sodium chloride, solar system, ***Solitary Sandpiper, Song Sparrow, Sora,*** species, speculative, *Spiral Jetty*, spiraling ecology, ***Spotted Sandpiper, Spotted Towhee,*** stars, steward, sticky, ***Stilt Sandpiper,*** stinky, stone, strata, stuck, subject, subjective, sublime, sulfur, *Sun Tunnels*, ***Surf Scoter,*** surface, surveying, survival, ***Swainson's Hawk, Swainson's Thrush,*** **about the power of microbial species, integral to our life cycles.** Talara Tar Seeps (Peru), Tanque Loma Tar Seep (Ecuador), tar balls, tar pits, tar sands (Canada), tar seeps, tar volcanoes, ***Thayer's Gull,*** time, timescale, ***Townsend's Solitaire,*** tracking, trails, travel, travelogue, ***Tree Swallow, Trumpeter Swan, Tundra Swan, Turkey Vulture,*** **Bacteria line our guts, maintain our body chemistry,** UGS (Utah Geological Survey), UMFA (Utah Museum of Fine Arts), understory, U.S. (United States), us, Utah, Ute, **and one day decompose us back to dirt—unless we get stuck in a tar seep. With the receding lake,** valley, value, vanishing, variation, ***Vaux's Swift, Vesper Sparrow,*** view, viewfinder, viewshed, ***Violet-green Swallow, Virginia Rail, Virginia's Warbler,*** virtual, viscous, vision, visit, vital, void, volatile, vulnerable, **tar seeps assert relations to reveal a 'dead sea' in the desert as deeply alive: a watershed for thinking about any overlooked place.** ***Warbling Vireo,*** Wasatch Mountains, water, water diversion, water rights, water runoff, watershed, wave, we, web, Weber River, well, Wendover, West, ***Western Grebe, Western Kingbird, Western Meadowlark, Western Sandpiper, Western Tanager, Western Wood-Pewee,*** wetland, ***Whimbrel, White-crowned Sparrow, White-faced Ibis, White-throated Sparrow, White-throated Swift, White-winged Scoter,*** wild, wildfire, wildlife, ***Willet, Willow Flycatcher, Wilson's Phalarope, Wilson's Snipe, Wilson's Warbler,*** wonder, ***Wood Duck,*** world, worldview, xeric, ***Yellow-headed Blackbird, Yellow-rumped Warbler, Yellow Warbler,*** you are here: **Revaluing what is dismissed—as *dying*, as *decomposing*, and derivatives—what might seep through to shift perceptions beyond the limits of language?**

Field Notes for Further Reading

a reverse stratigraphy

Many sources about Great Salt Lake have been consulted to round out its complex dynamics: scientific, historic, cultural, economic, artistic, and other qualities. The following notes toward further reading only scrape the surface and relate primarily to quoted material. In addition to published background resources, web resources have been consulted from the official sites for the Great Salt Lake Institute, Utah Museum of Fine Arts, Dia Art Foundation, Utah Department of Natural Resources, Bear River Migratory Bird Refuge, U.S. Fish & Wildlife Service, Utah Geological Survey, Bureau of Land Management, National Park Service, Holt/Smithson Foundation, Land Arts of the American West, Center for Land Use Interpretation, *Salt Lake City Tribune,* and related agencies and organizations, along with contemporaneous news and historical reports. • In most instances throughout the book, I deleted the article "the" before Great Salt Lake, since locals drop the article. As Jaimi Butler explained, "We don't say 'the Utah Lake' or 'the Lake Ontario.'" • The book's epigraphs are excerpted from Robert Smithson, "A Sedimentation of the Mind: Earth Projects" (1968), in *Robert Smithson: The Collected Writings*, ed. Jack Flam (Berkeley, CA: University of California Press, 1996), pp. 102, 110; Aldo Leopold, *A Sand County Almanac:*

Sketches Here and There (Oxford: Oxford University Press, 1989 [1949]), p. 159; Emily Dickinson, (46b) "'Hope' is the thing with feathers—," in *The Complete Poems of Emily Dickinson*, ed. Thomas H. Johnson (Boston: Little, Brown & Co., 1951), p. 254; and Rachel Carson, *Silent Spring* (Boston: Houghton Mifflin, 1962), p. 8. Epigraphs for chapters are noted in corresponding sections.

PREFACE & PROLOGUE: *composition & decomposition*

The definition of "ecotone" comes from *Ecotone*, which published an excerpt from "Life in the Tar Seeps" (see Acknowledgments). • Aldo Leopold, *A Sand County Almanac: Sketches Here and There* (Oxford: Oxford University Press, 1989 [1949]), p. 130. • Robin Wall Kimmerer, "Nature Needs a New Pronoun: To Stop the Age of Extinction, Let's Start by Ditching 'It'," *YESMagazine.org*, 03/30/2015 • Maria Sibylla Merian, *Mariæ Sibillæ Merian Dissertatio de generatione et metamorphosibus insectorum surinamensium* (Amstelædami: Apud J. Oosterwyk, 1719), n.p., plate 9. Courtesy of the John Carter Brown Library. • Ann Reynolds, "The Problem of Return," in *Land Arts of the American West*, eds. Chris Taylor and Bill Gilbert (Austin, TX: University of Texas Press, 2009), p. 129. • Nancy Holt, "Sun Tunnels" (first published in *Artforum* in 1977), quoted in Alena J. Williams, *Nancy Holt: Sightlines*, ed. Alena J. Williams (Berkeley, CA: University of California Press, 2015), p. 34. • Jaimi Butler, personal conversation, quoted in "Unspiraling."

I. GREAT SALT LAKE: *Death Traps*

Epigraph for this section comes from Terry Tempest Williams, *Refuge: An Unnatural History of Time and Place* (New York: Vintage, 1992 [1991]), p. 149. • My first visit to Rozel Point occurred on October 15, 2017, bypassing the tar seeps for *Spiral Jetty*; this chapter follows my second trip on February 12, 2018 (coincidentally the birthday of Charles Darwin) when I first visited the tar seeps. • To visit Rozel Point at Great Salt Lake, guides are available, including from the Utah Museum of Fine Arts in a pamphlet titled "Robert Smithson, *Spiral Jetty*, 1970: Experiential Guide." For a geological field guided tour of the region, see Mark Milligan and H. Gregory McDonald, "Shorelines and vertebrate fauna of Pleistocene Lake Bonneville, Utah, Idaho, and Nevada," *Geology of the Intermountain West*, v.4 (2017): pp. 181–214. For background about the paleontology of tar pits and seeps, see H. Gregory McDonald, John M. Harris, and Emily Lindsey, "Introduction," *La Brea and Beyond: The Paleontology of Asphalt-Preserved Biotas*, ed. John M. Harris (Los Angeles, CA: Natural

History Museum of Los Angeles, 2015), pp. 1–4. For a scientific overview of Great Salt Lake, see *Great Salt Lake Biology: A Terminal Lake in a Time of Change*, eds. Bonnie K. Baxter and Jaimi K. Butler (Cham, Switzerland: Springer International, 2020). For background about halophiles and microbial studies of Great Salt Lake, a few articles include Bonnie K. Baxter, "Great Salt Lake Microbiology: A Historical Perspective," *International Microbiology* (04 June 2018); Daniel L. Jones and Bonnie K. Baxter, "DNA Repair and Photoprotection: Mechanisms of Overcoming Environmental Ultraviolet Radiation Exposure in Halophilic Archaea, *Frontiers in Microbiology* (29 September 2017); Rachel Buchanan, "Utah microbes point to Mars," *BBC News* (18 May 2004). • Regarding Indigenous history in Utah, some background can be found in Steven R. Simms, *Ancient Peoples of the Great Basin and Colorado Plateau* (Walnut Creek, CA: Left Coast Press, 2008); Jared Farmer, *On Zion's Mount: Mormons, Indians, and the American Landscape* (Cambridge, MA: Harvard University Press, 2008); David Rich Lewis, "Native Americans in Utah," *Utah History Encyclopedia*, ed. Allan Kent Powell (Salt Lake City, UT: University of Utah Press, 1994); Hikmet Sidney Loe, "American Indians," in *The Spiral Jetty Encyclo* (Salt Lake City, UT: University of Utah Press, 2017), pp. 37–43; and widening into the Southwest through works including Stephen Trimble, *The People: Indians of the American Southwest* (Santa Fe, NM: SAR Press, 1993) and *Red Rock Stories: Three Generations of Writers Speak on Behalf of Utah's Public Lands* (Salt Lake City, UT: Torrey House Press, 2017). Early recorded Goshute names for the lake are noted in Ralph V. Chamberlin, "Place and Personal Names of the Gosiute Indians of Utah," *Proceedings of the American Philosophical Society* 52/208 (1913): 1–20. Note that these sources are primarily White accounts of Native American histories. • Sources for early Western expeditions include Howard Stansbury, *Exploration and Survey of the Valley of the Great Salt Lake of Utah* (Philadelphia, PA: Lippincott, Grambo & Co., 1852), pp. 101–2; Howard Stansbury, ed. Brigham D. Madsen, *Exploring the Great Salt Lake: The Stansbury Expedition of 1849–50* (Salt Lake City, UT: University of Utah Press, 1989 [1855]), p. 491; and John Charles Fremont, *Report of the exploring expedition to the Rocky Mountains . . .* (Washington: Gales and Seaton, 1845); "Valle Salado" is noted on a map in Sidney E. Morse, *New Universal Atlas of the World on an Improved Plan of Alphabetical Indexes, Designed for Academies and Higher Schools* (1826), reproduced in Stephanie Earls, "Historical Maps: More than Meets the Eye," *Utah Geological Survey Notes* (May 2011), pp. 5–7, 11; for earlier history, see John L. Kessell, *Whither the Waters: Mapping the Great Basin From Bernardo de Miera to John C. Frémont* (Albuquerque, NM: University of New Mexico Press, 2017). • Additional works about the artist and earthwork include

Robert Smithson, "The Spiral Jetty" (1972), in *Robert Smithson Spiral Jetty: True Fictions, False Realities*, eds. Lynne Cooke and Karen Kelly (Berkeley, CA: University of California Press, 2005), pp. 7–13; *Robert Smithson*, eds. Eugenie Tsai and Cornelia Butler (Los Angeles, CA: Museum of Contemporary Art, 2004); Ann Reynolds, *Robert Smithson: Learning from New Jersey and Elsewhere* (Cambridge, MA: MIT Press, 2003); Jennifer L. Roberts, *Mirror-Travels: Robert Smithson and History* (New Haven, CT: Yale University Press, 2004). • Many writers have questioned narrative strategies around environmental changes and the climate crisis: from Amitov Ghosh, *The Great Derangement: Climate Change and the Unthinkable* (Chicago, IL: University of Chicago Press, 2017); to Ursula Heise, *Imagining Extinction: The Cultural Meanings of Endangered Species* (Chicago, IL: University of Chicago Press, 2016); to Robin Wall Kimmerer, *Braiding Sweetgrass: Indigenous Wisdom, Scientific Knowledge, and the Teachings of Plants* (Minneapolis, MN: Milkweed, 2015), to Brendan Larson, *Metaphors for Environmental Sustainability: Redefining Our Relationship with Nature* (New Haven, CT: Yale University Press, 2011); to Robin Kundis Craig, "Learning to Live with the Trickster: Narrating Climate Change and the Value of Resilience Thinking," *Pace Environmental Law Review* 33 (Spring 2016), pp. 351–396; to Siri Veland, et al., "Narrative matters for sustainability: the transformative role of storytelling in realizing 1.5°C futures," *Current Opinion in Environmental Sustainability* 31 (2018): 41–47; to name a few. See also LeAnne Howe, "The Story of America: A Tribalography," in *Choctalking on Other Realities* (San Francisco, CA: Aunt Lute Books, 2013), pp. 13–40; and Diana James, "*Tjukurpa* Time," *Long History, Deep Time: Deepening Histories of Place*, eds. Ann McGrath and Mary Anne Jebb (Canberra: Australian National University Press, 2015), pp. 33–45. • Fredric (Fritz) Norstad, "A Different Take on Ecology," *Holden Village Archive* (Chelan, WA: 1986). • Louise Erdrich quoted in Peter Beidler & Gay Barton, *A Reader's Guide to the Novels of Louise Erdrich* (Columbia, MO: University of Missouri Press, 2006), p. 381. • Faded word lists around *Spiral Jetty* quoted from Spencer Finch's installation *Great Salt Lake and Vicinity* (2017) in the Utah Museum of Fine Arts located in Salt Lake City, Utah. Visiting Finch's artwork in 2018, I was struck by the lack of represented birds, tar seeps, and people. Read more in the margins of the next chapter. • For background about extremophiles and medical research, a few sources include: Prasanti Babu, Anuj Chandel, and Om Singh, eds., *Extremophiles and their Applications in Medical Processes* (Cham, Switzerland: Springer, 2015), pp. 31–32; and Dinesh Maheshwari and Meenu Saraf, eds., *Halophiles: Biodiversity and Sustainable Exploitation* (Cham, Switzerland: Springer, 2015), p. 292. • John Coplans quoted in Robert Hobbs, *Robert Smithson: Sculpture* (Ithaca, NY: Cornell University Press, 1981), p. 47; see also Ric Collier and Jim Edwards, "Spiral Jetty: The Re-Emergence," *Sculpture* (July/August 2004). • Stephen Hawking, *A Brief History of Time* (New York: Bantam, 1988), pp. 144–145. • For symbolism of varied birds, see Desmond Morris, *Owl* (London: Reaktion Books, 2009) and Barbara Allen, *Pelican* (London: Reaktion Books, 2019), among others. • United States, *Great Salt Lake National Park in Utah: hearings before the Subcommittee on Public Lands of the Committee on Interior and Insular Affairs, United States Senate, Eighty-sixth Congress, Second session, on S. 2894, a bill to authorize the Secretary of the Interior to establish the Great Salt Lake National Park in the state of Utah, Salt Lake City, Utah, November 10, 1960, Ogden, Utah, November 12, 1960* (1961).

II. AMERICAN WEST: *Stuck*

Epigraph from Susan Sontag, *On Photography* (New York: Penguin, 1977), p. 15. • To explore representations of land through the consequences of colonial-settlement, a few sources include Alan Trachtenberg, "Naming the View," in *Reading American Photographs: Images as History, Mathew Brady to Walker Evans* (New York: Farrar, Straus and Giroux, 1990), p. 119–163; William Cronin, *Uncommon Ground: Rethinking the Human Place in Nature* (New York: W.W. Norton, 1996); Denise D. Meringolo, "Managing the Landscape: National Parks, National Monuments, and the Use of Public Land," in *Museums, Monuments, and National Parks: Toward a New Genealogy of Public History* (Amherst, MA: University of Massachusetts Press, 2012), pp. 26–55; Ariella Azoulay, *Potential History: Unlearning Imperialism* (New York: Verso, 2019); Carolyn Finney, *Black Faces, White Spaces: Reimagining the Relationship of African Americans to the Great Outdoors* (Chapel Hill, NC: University of North Carolina Press, 2014); Lauret Savoy, *Trace: Memory, History, Race, and the American Landscape* (Berkeley, CA: Counterpoint, 2016), quoted in Acknowledgments; among many others. In *Everything You Know About Indians Is Wrong*, Paul Chaat Smith (Comanche) addresses dangers in reductive colonial-settler concepts of "land," "place," "time," and "history," adding: "We come from a different place" shaped by the ever-present "land question," and "no history is complete without knowing the history of the history" (Minneapolis, MN: University of Minnesota Press, 2009, pp. 53, 85). Historical concepts of humans as separate from environments have enforced exclusions from nature and wilderness along lines of race, gender, age, disability, and other markers of difference, with repercussions for new generations; as only one testimonial, see Lauren McGrady, "My Present Is Not Your Tombstone," in *Coming of Age at the End of Nature: A Generation Faces Living on a Changed Planet*, eds. Julie Dunlap and Susan A. Cohen (San Antonio,

TX: Trinity University Press, 2016), pp. 109–110. • Great Salt Lake has inspired a range of artistic works and cultural critiques connected with the wider American West; see websites for and works by the Center for Land Use Interpretation, Land Arts of the American West, Terminal Lake Exploration Platform (formerly Great Salt Lake Exploration Platform), and websites and works by artists including Motoi Yamamoto (a visiting artist at the Great Salt Lake Institute in 2014) and the interdisciplinary arts collective Postcommodity (Cristóbal Martínez, Kade L. Twist, Raven Chacon), including the documentary film *Through the Repellent Fence*, directed by Sam Wainwright Douglas (2017), along with Emily Eliza Scott, "Decentering Land Art from the Borderlands: A Review of *Through the Repellent Fence*," *Art Journal Open* (March 27, 2018). Further considerations arise by contrasting Land Art with Native American earthworks; see also Eric Gary Anderson, "Earthworks and Contemporary Indigenous American Literature: Foundations and Futures," *Native South* 9 (2016), pp. 1–24. • The consideration of *Spiral Jetty* as a "decomposing platform" climatically adapts Ann Reynold's concept of "image/platform" in "Casting Glances: Reconsidering Smithson's 'Documentary' Process," in *Art in the Landscape* (Marfa: Chinati Foundation, 2000), pp. 55–71. I am grateful to Ann Reynolds and Chris Taylor for describing their visits to Rozel Point before much oil drilling equipment was removed. • Myriad sources convey environmental changes over time at Rozel Point, along with Smithson's engagement with asphalt and tar, and broader intersections with the Land Arts movement. Some include Robert Smithson, "The Spiral Jetty" (1972) and "A Sedimentation of the Mind: Earth Projects" (1968), *Robert Smithson: The Collected Writings*, pp. 100–113, 143–153; Hikmet Sidney Loe, "Asphalt" and "Land Reclamation," *Spiral Jetty Encyclo*, pp. 45–47, 154; Robert Smithson, "*Spiral Jetty:* The Film (1970)," transcribed and annotated by Serge Paul, in Loe, *Spiral Jetty Encyclo*, pp. 22–25; Hikmet Sidney Loe, "A Rediscovered Nancy Holt Documentary from 1970," *Hyperallergic* (03/05/2020); Leslie Ryan, "Art + Ecology: Land Reclamation Works of Artists Robert Smithson, Robert Morris, and Helen Mayer Harrison and Newton Harrison," *Environmental Philosophy* 4/1&2 (2007), pp. 95–116; Mark Milligan, "Geosights: Rozel Point and Spiral Jetty Revisited, Box Elder County, Utah," *Utah Geological Survey: Survey Notes*, v. 38/2 (May 2006); Robert Smithson, *Asphalt Rundown* (1969), Holt/Smithson Foundation online, along with commentaries on both Smithson and Holt by Lisa Le Feuvre; Laura Raicovich quoted in Kirk Johnson, "Plans to Mix Oil Drilling and Art Clash in Utah," *New York Times* (03/27/2008); Dia Art Foundation, "Dia Art Foundation Announces Collaboration with the Great Salt Lake Institute and the Utah Museum of Fine Arts" (February 2, 2012); Andrew Menard, "Robert Smithson's Environmental History," *Oxford Art Journal* 37/3

(December 2014): 285–304; James Crump, *Troublemakers: The Story of Land Art* [Nonstop Entertainment], 2017; Chris Taylor, "Troubling *Troublemakers*," *Art Journal* 75/2 (Spring 2016): pp. 88–91; Chris Taylor and Bill Gilbert, *Land Arts of the American West* (Austin, TX: University of Texas, 2009); Suzaan Boettger, *Earthworks: Art and the Landscape of the Sixties* (Berkeley, CA: University of California Press, 2002); John Beardsley, *Earthworks and Beyond: Contemporary Art in the Landscape* (New York: Abbeville Press, 1984). References to "site" derive in part from Smithson's distinction between "sites" and "non-sites" as an interplay between outdoors and indoors, natural and built, dispersed and contained, concrete and abstract, displaced and placed, there and here. For Smithson, the "site" was "the physical, raw reality" of a location while the "non-site" represented a sampling of that reality elsewhere. Nancy Holt set up further "sightlines." Writing about Smithson's "Sites/Nonsites," Lawrence Alloway wrote: "The relation of the Nonsite to the Site is also like that of language to the world." It might be said that my attempt to write this book about Rozel Point's tar seeps is a non-site of that site with sightlines. • Considering entangled aesthetic, cultural, and environmental representations, see Gretchen E. Henderson, *Ugliness: A Cultural History* (London: Reaktion Books, 2015), p. 11, 164. • After the camera traps were set in April 2018 (led by Kara Kornhauser '19 of Westminster College), the first animals to get stuck were a gull and a snake. Kara helped to deduce that the "pelican death assemblage" was caused by an uncapped oil well, which led to its capping in 2019, and as noted earlier, she co-wrote a chapter with Greg and Jaimi on "Rozel Point Tar Seeps and Their Impact on the Local Biology at Great Salt Lake, Utah" in *Great Salt Lake Biology: A Terminal Lake in a Time of Change.* • Among other works related to environmental justice and extractions of natural resources, see Rob Nixon, *Slow Violence and the Environmentalism of the Poor* (Cambridge, MA: Harvard University Press, 2011).

III. NORTH AMERICA: *Unspiraling*

Epigraph by Donna J. Haraway, *Staying with the Trouble: Making Kin in the Chthulucene* (Durham, NC: Duke University Press, 2016), p. 1. • QR codes in this chapter are explained in appendices on "Field Notes on Looking: A Visual Itinerary" and "Field Notes on Method: Fossils in the Making." • Marginal alphabetized lists in this chapter (inspired by Spencer Finch's *Great Salt Lake and Vicinity*) expand vocabulary around this story akin to a fragmented index that is otherwise missing from its traditional place at the end of the book, suggesting declassification; see appendix "Unindexing: Seeping through the Limits of Classification." • Quotations include Robert Smithson, "The Spiral

Jetty," *Arts of the Environment*, ed. Gyorgy Kepes (George Braziller, 1972), pp. 223, 227, 229 (includes internal quotation from Thomas H. Clark, Colin W. Stern, *Geological Evolution of North America*, n.d., New York, Ronald Press Co., p. 5). • Some works of Land Art alluded to are made by Hamish Fulton, Richard Long, Andy Goldsworthy, James Turrell, Richard Serra, Walter De Maria, Michael Heizer, Nancy Holt, Postcommodity, Robert Smithson, Spencer Finch, and Motoi Yamamoto. • Bryan Wagner, *The Tar Baby: A Global History* (Princeton, NJ: Princeton University Press, 2017). • The presentation by the Spiral Jetty Partnership took place at the Art Libraries Society of North America Conference on March 29, 2019 in Salt Lake City, UT, including Bonnie Baxter (Director, Great Salt Lake Institute at Westminster College); Whitney Tassie (Senior Curator, Curator of Modern and Contemporary Art, Utah Museum of Fine Arts); Laura Ault (Sovereign Lands Program Manager, Division of Forestry, Fire and State Lands); and Kelly Kivland (Associate Curator, Dia Art Foundation). Among other sources, see Jeffrey Kastner, "Entropy and the New Monument: On the Future of *Spiral Jetty*," *Artforum* (April 2008): 167–70. • Sally Jewell in discussion at University of Utah Law School for symposium, "Recreation Challenges on Public Lands," March 21, 2019. • "The Price of Nature" was discussed during a talk by Udo Weilacher at Dumbarton Oaks in Washington, D.C.: "Between Land Art and Landscape Architecture: A Dialogue with Udo Weilacher and John Beardsley" (30 March 2017). • Greg Saris (Miwok) quoted in LeAnne Howe, "The Story of America: A Tribalography," *Choctalking on Other Realities* (San Francisco, CA: Aunt Lute Books, 2013), pp. 13–40. • Kathryn Schultz, "The Really Big One," *New Yorker* (20 July 2015). • Sherry Turkle and Tim Adams, "I am not anti-technology, I am pro-conversation," *Guardian,* 18 October 2015. • Christopher D. Stone, "Should Trees Have Standing—Toward Legal Rights for Natural Objects," *Southern California Law Review* 45 (1972): 450–501. • For background about seismic studies of red rock arches, see Katherine Kornie, "Are these arches going to collapse? Meet the scientist who's trying to find out," *Science* (28 July 2017); Cheryl Dybas and Paul Gabrielsen, "Song of the Red Rock Arches," *National Science Foundation* (5 December 2017); Jeff Moore and Paul Gabrielsen, "Resonance in Rainbow Bridge," *University of Utah UNews* (21 September 2016); and of microseisms in Great Salt Lake, see Keith Koper, "Waves in Lakes Make Waves in the Earth," *University of Utah UNews* (16 October 2017). Thanks to the National Park Service and the Native American Consultation Committee at Rainbow Bridge for including me with geologists Jeffrey Moore and Riley Finnegan in their 2019 annual meeting. • Quotes from 1896 land deed in Utah; letter from William Harroun Behle Collection, "Bird Sanctuary Documents 1934–1957," in Special Collections, J. Willard Marriott Library,

University of Utah; and Carina Wyborn, et al., "Imagining transformative biodiversity futures," *Nature Sustainability*, 3/9 (2020): 670–672. • Quote from Hua Hsu, "The Search for New Words to Make Us Care about the Climate Crisis," *New Yorker* (02/21/2020), reviewing *An Ecotopian Lexicon*, eds. Matthew Schneider-Mayerson and Brent Ryan Bellamy (Minneapolis, MN: University of Minnesota Press, 2019). • Thanks to family **김영혜** for teaching me the Korean word for **물수리**.

EARTH: *The Big Here*

Epigraph from Ursula Heise, *Imagining Extinction: The Cultural Meanings of Endangered Species* (Chicago: University of Chicago Press, 2016), p. 12. • "The Big Here" is an adapted excerpt from my essay on "Intermedia Genres: Breathing Lessons in Changing Climates," published in *Notre Dame Review* (Fall 2019/Winter 2020). Some of the sources that influenced this section include Juhani Pallasmaa, *The Eyes of the Skin: Architecture and the Senses* (Chichester, West Sussex: Wiley & Sons, 2005); David Abrams, *Becoming Animal: An Earthly Cosmology* (New York: Vintage, 2011); Edward Rothstein, "Extreme Museum: The Rigors of Contemplation," *New York Times* (October 21, 2011); Jourdan Imani Keith, "Desegregating Wilderness," *Orion* (September 10, 2014); William McDonough, "Carbon is Not the Enemy," *Nature: International Weekly Journal of Science* 539/7629 (15 November 2016); Robin Wall Kimmerer, *Braiding Sweetgrass: Indigenous Wisdom, Scientific Knowledge, and the Teachings of Plants* (Minneapolis, MN: Milkweed, 2015); Eduardo Kohn, *How Forests Think: Toward an Anthropology Beyond the Human* (Berkeley, CA: University of California Press, 2013); Jeffrey Jerome Cohen and Lowell Duckert, eds., *Veer Ecology: A Companion for Environmental Thinking* (Minneapolis, MN: University of Minnesota Press, 2017); Richard Powers, *The Overstory* (New York: W.W. Norton & Co., 2018); and many iterations of "Once upon a time." • For the "Coda," Nora Naranja-Morse's poem on "Always Becoming" is quoted from D.B. Spruce and T. Thrasher, eds., *The Land has Memory: Indigenous Knowledge, Native Landscape, and the National Museum of the American Indian* (Chapel Hill, CA: University of North Carolina Press, 2009), pp. 61–70.

FIELD NOTES: *excavations*

The epigraph for "Field Notes: Excavations" comes from Rebecca Solnit, *A Field Guide to Getting Lost* (New York: Penguin, 2005), p. 6. • **Field Notes on Water: Runoff from Your Watershed** is adapted and expanded from Peter Warshall's

Watershed Awareness Exercise (*CoEvolution Quarterly*, Winter 1976–1977) that has evolved through others over the years into variations of "The Big Here Quiz" (including an online edition by Kevin Kelly, 2005); for epigraph with further reading, refer to Peter Warshall, "Water Governance: Checklists to Encourage Respect for Waterflows and People," in *Writing on Water*, eds. David Rothenberg and Marta Ulvaeus (Cambridge: MIT Press, 2002), p. 46. Note: I have offered 52 questions for 52 weeks: as a water reflection for the year ahead. • **Field Notes on Looking: A Visual Itinerary** includes epigraph by Lucy R. Lippard, from *Undermining: A Wild Ride through Land Use, Politics, and Art in the Changing West* (New York: The New Press, 2014), p. 176. • **Field Notes on Method: Fossils in the Making** (on material-digital seepage): For additional readings & resources, follow the QR codes or proceed here: https://www.gretchenhenderson.com/life-in-the-tar-seeps. For cursory background, see Gretchen E. Henderson, *Galerie de Difformité* (Chicago, IL: &NOW Books: Northwestern University Press, 2011) and "This Is ~~Not~~ a Book: Melting Across Bounds," delivered as keynote address for an exhibition at Hampshire College/Five Colleges, MA, on "Pulp to Pixels: Artists' Books in the Digital Age" (2012) and published in *Journal of Artists' Books*, 33 (Spring 2013): pp. 29–33. • Epigraphs for **UnIndexing: Seeping through the Limits of Classification** include Lyanda Lynn Haupt, *Rooted: Life at the Crossroads of Nature, Science, and Spirit* (New York: Little, Brown, Spark, 2020), p. 152; Ralph Waldo Emerson, "The Poet," quoted in "Nature's Archive" by Eduardo Cadava, *Emerson and the Climates of History* (Redwood City, CA: Stanford University Press, 1997), p. 93, and Robert Smithson, "The Spiral Jetty," *The Collected Writings*, ed. Jack Flam (Berkeley, CA: University of California Press, 1996), p. 149. Wendell Berry also describes "the word is a fossil" in *The Unsettling of America: Culture and Agriculture* (Berkeley, CA: Counterpoint, 2015 [1977]), p. 128. Integrated names of birds come from the U.S. Fish & Wildlife Service's "Bird List" for the Bear River Migratory Bird Refuge, published in March 2006 and as of 2019 available for download on the refuge's website.

Below is an intermedia epigraph by Tim Ingold from *Being Alive: Essays on Movement, Knowledge and Description* (New York: Routledge, 2011), p. xii:

Why do we acknowledge only our textual sources but not the ground we walk, the ever-changing skies, mountains and rivers, rocks and trees, the houses we inhabit and the tools we use, not to mention the innumerable companions, both non-human animals and fellow humans, with which and with whom we share our lives?

Each of us, too, is a landscape inscribed . . .
—Lauret Savoy, *Trace*

"Field Note: Jaimi Butler holding pelican tracking tags," Rozel Point, Great Salt Lake, Utah, 02/12/2018.

"Field Note: Greg McDonald calibrating GPS for tar seep with barn owls," Rozel Point, Great Salt Lake, Utah, 02/12/2018.

Acknowledgments

For a book inspired by collaborative environmental stewardship, there are many to thank: humans and beyond. More could have grown this book beyond these pages. The following notes attempt to value that breadth while acknowledging many limits amid wider ongoing efforts by many to steward Great Salt Lake.

This spiraling essay arose from an invitation from Jaimi K. Butler, coordinator of the Great Salt Lake Institute, to accompany her and H. Gregory McDonald, regional paleontologist for the Bureau of Land Management, to the tar seeps beside Robert Smithson's *Spiral Jetty* (1970). I remain grateful to Jaimi and Greg for welcoming me to Great Salt Lake in February 2018, for sharing their experience and enthusiasm through multiple field trips to Rozel Point and for ensuing conversations. Thanks also to Bonnie Baxter, founder and director of the Great Salt Lake Institute at Westminster College, who for two decades has supported myriad collaborations on behalf of the Lake. Along the way, I had the good fortune to meet a variety of the GSLI's environmental and artistic collaborators and student scientists, including Kara Kornhauser '19 of Westminster College who led student research of Rozel Point's tar seeps from summer 2018 to summer 2019. The growing commitment by new generations embodies care and hope.

In many ways the book's ending doubles as a beginning. This book was largely written by accident, and its limitations are my own. I came to Utah, in part, to pick up an earlier project of "A Philosophy of Stones" that ironically had scattered after otoliths (stones in my inner ear) dislodged when I was hit by a car in a crosswalk. While following cairns, instead I got stuck in tar seeps.

For a book grounded in deep time, deep thanks are owed. I am grateful to the Annie Clark Tanner Fellowship in Environmental Humanities at the University of Utah that supported my research, writing, and teaching. Thanks to colleagues in the Tanner Humanities Center and in Environmental Humanities, including Bob Goldberg, Jeff McCarthy, and Julia Corbett for the fellowship invitation and connection to their programs. Special thanks to Susan Anderson who introduced me to the GSLI and was an early and ongoing supporter of this book. I appreciated invitations to present excerpts and photographs from this project at the Taft-Nicholson Center for Environmental Humanities in Montana (June 2018) and at the Tanner Humanities Center at the University of Utah (February 2019), also providing me with opportunities to blend my earlier work on aesthetics of ugliness and beauty into environmental studies and narratives of the climate crisis. The gift of an artist residency at the Taft-Nicholson Center widened my sense of the West, bird and other migrations, and collaborative stewardship. In varied venues crossing urban and wild terrains, I was grateful for feedback from listeners and readers, including Susan Anderson, Mark and Carol Bergstrom, Bob Goldberg, Heather Houser, Carlos Santana, Claudia Esslinger, among others, including conversations with Whitney Tassie at the Utah Museum of Fine Arts, Hikmet Sidney Loe, Trent Alvey, Jeffrey Moore, and a range of guests who visited my courses and hosted site visits that locally grounded my interdisciplinary and pedagogical work. My gratitude extends to students in my varied graduate courses ("Tectonic Essays," "Writing as Archaeology," "Environmental Writing in the Digital Age") for their commitments to environmental stewardship and social justice. I appreciated everyone who welcomed me to various communities around Salt Lake City and farther afield.

Thanks to editor Marguerite Avery who believed in this book and to anonymous reviewers who provided valued feedback. Excerpts from this book were published in different forms in *Ecotone* (as "Life in the Tar Seeps," Winter 2019/2020), *Ploughshares* (as "Thinking Like a Crosswalk," Winter 2019/2020), *Notre Dame Review* (as "Intermedia Genres: Breathing Lessons in Changing Climates," Winter 2019/2020), and *Seeds of Change: Provocations for a New Research Agenda on Biodiversity Revisited* by the Luc Hoffmann Institute/World Wildlife Fund International (as "Listen for a Pelican, Owl, Gull, Hawk, and Chickadee: Narratives for Biodiversity Revisited," 2019). My gratitude extends to Anna Lena Phillips Bell, Elliot Emory Smith, Ladette Randolph, Steve Tomasula, Carina Wyborn, and editorial teams of these journals and proceedings who selected or invited my work for publication. Thanks also to Duke University's Center for Documentary Studies that commended an excerpt of this project in 2019 and to the Interdisciplinary Humanities Center at the University of California at Santa Barbara that selected some photographs for their 2019–2020 Platform Gallery exhibition on *Critical Mass.* The questions asked in the book's "Field Notes on Listening" echo questions included in my essay on "Sharing and Shaping Space: Notes toward an Aesthetic Ecology," in *Interdisciplinary Approaches to Disability: Looking Towards the Future,* eds. Katie Ellis, Rosemarie Garland-Thomson, Mike Kent, Rachel Robertson (London: Routledge, 2019). I remain grateful for the support and community of the John Carter Brown Library at Brown University and the Hodson Trust–JCB Fellowship that supported my research residency (winter/spring of 2016) where I saw Maria Sybila Merian's rare book on metamorphosis and, in a nearby arts library, a photograph of Nancy Holt's *Sun Tunnels.* Thanks also to the Jan Michalski Foundation for Writing and Literature for a fellowship residency in nature writing (May 2019) in Montricher, Switzerland, where Jessica Villat invited me to collaborate with the Luc Hoffmann Institute/World Wildlife Fund International to share part of this story. The Biodiversity Revisited initiative and conversations led me outside this book to convenings in Vienna, Austria, and the Rockefeller Foundation's Bellagio Center at Lake Como in Italy. Thanks to those who I met around the globe to collaborate on co-authoring articles for *Nature Sustainability* ("Imagining transformative biodiversity futures," 2020) and *Conservation Biology* ("An agenda for research and action towards diverse and just futures for life on Earth," 2020). The process was more than the product, planting seeds to water and grow.

As Great Salt Lake is a living watershed, more runoff will stick to these pages through QR codes as material and digital tributaries, where gratitude will be extended virtually in situ.

In keeping with Acknowledgments, I (We) wish to echo the University of Utah's Land Acknowledgement (2020) while acknowledging that and my own and this book's shortcomings:

Given that the Salt Lake Valley has always been a gathering place for Indigenous peoples, I (We) acknowledge that this land, named for the Ute Tribe, is the traditional and ancestral homelands of the Shoshone, Paiute, Goshute, and Ute Tribes and is a crossroad for Indigenous peoples. Situated within a network of historical and contemporary relationships, I (We) recognize the enduring relationships between many Indigenous peoples and their traditional homelands and respect that Utah's Indigenous peoples have been the original stewards of this land. I (We) value the sovereign relationships that exist between tribal governments, state governments, and the federal government. Today, approximately 60,000 American Indian and Alaskan Native peoples live in Utah. I (We) also recognize that any acknowledgment is not ongoing action and is only a small step toward cultivating strong relationships with Indigenous communities.

Since this book is being published by a publisher based in Texas, I also wish to echo a proposed Land Acknowledgment where I recently started working at the University of Texas at Austin (2020), with thanks to those who articulated: We (I) would like to acknowledge that we are meeting on the Indigenous lands of Turtle Island, the ancestral name for what now is called North America. Moreover, (I) We would like to acknowledge the Alabama-Coushatta, Caddo, Carrizo/Comecrudo, Coahuiltecan, Comanche, Kickapoo, Lipan Apache, Tonkawa and Ysleta Del Sur Pueblo, and all the American Indian and Indigenous Peoples and communities who have been or have become a part of these lands and territories in Texas.

Considering this book's digital extension, I also wish to add an acknowledgment for reading across platforms (echoing one written by Amy Smith and Emily Johnson and shared by Creative Capital): As we gather in the digital realm, let us address and change inequitable access to technology and environmental and other injustices embedded in the making and distribution of the digital technologies that we use. Let us acknowledge our responsibility to each other and to the land that we occupy, and move toward reevaluating and repairing our commitments in every space, including the digital.

As I have lived across many geographies, I am indebted to many more lands and lives.

Deep gratitude to friends and family near and far, to the Ernsters and Norstads and all of my forebears, for your complex histories and for teaching me to question and to love the world deeply and daily. Special gratitude to my parents, Martin and Virginia Ernster. For my entire life, I have felt fortunate to be able to spend time on this vulnerable and volatile planet, as it changes and teaches me to change. There are always teachers to thank, and some of my early ones, including John McPhee and Will Howarth, left indelible impressions around my interrelated senses of literature and environment. Thanks to many educational settings around the country that have enabled me to engage students over many years with various forms of field work, spiraling from my early forays in the Sierra Nevada. Sierra reminds me to love unconditionally in the present tense. Thanks always to my husband, Ethan, who has co-created home with me in many places and who knows what I cannot articulate. My gratitude extends to all of these and many more: past, present, future. Whether named here or not, known or unknown, I am grateful to you. Thank you for accompanying me at this moment, in these pages and on Earth, here and now.

TOGETHER:

Thanks to the ground we walk. To the ever-changing skies. To rivers and mountains, rocks and trees. To our habitations that we inhabit like snail shells, which at a certain point may no longer fit, and to tools that we use and negotiate as if appendages of our bodies. May we learn to adapt our tools and habitations respectfully and, when they get out of hand and damage the land of which we are part, may we recalibrate and retreat with humility for a restorative and healing balance. Thanks to innumerable companions, both fellow humans and non-human animals. When we inhabit an integrated body of mind and heart, we are extensions of each other and can work to support each other's health and wellbeing. Forgive whatever habits I may have learned that may separate us; I continue to learn and unlearn at once. This process of learning and unlearning will occur as long as I live. Just as the Earth is alive and changing, I hope that we can continue to change together to support reciprocity, resilience, and renewal. Thank you for letting me share your lives in the small passage of this book.

YOU ARE HERE.

praise to the tar seeps
sticking together
matters: this water, this
art, tar, mud, sky, brine,
salt, sea, rocks, birds
feathering into focus:
wingbeats and wind,
sand crunches
underfoot
all matters:
all waters
retreat, inscribing
—lake level
—lake level
—lake level
revealing
less snow
more seeps
where language eludes
(*disarticulated:* bones,
words) getting stuck
in "death traps"
as a "dead sea" lives
as a body
of water, of land
among bodies
(human, animal, botanical)
interacting as mineral
extract-
ions seep
into our cells,
infinitesimal spirals
transforming
as we sing: praise

Gretchen Ernster Henderson writes across environmental arts, cultural histories, and integrative sciences. Her recent essays have appeared in *Ecotone*, *Ploughshares*, and the *Kenyon Review*, with co-authored articles in *Nature Sustainability* and *Conservation Biology*. Her four previous books include *Ugliness: A Cultural History* and *Galerie de Difformité*, cross-pollinating genres and arts and translated across five languages. She is a senior lecturer at the University of Texas at Austin and has also taught at Georgetown University, MIT, and the University of Utah, where she was the 2018–19 Annie Clark Tanner Fellow in Environmental Humanities. Born and raised in California, she is the 2023 Aldo and Estella Leopold Writer in Residence in New Mexico and lives in Arizona.

Published by Terra Firma Books, an imprint of Trinity University Press
San Antonio, Texas 78212

Cover design by Derek Thornton / Notch Design
Book design by Anne Richmond Boston

978-1-59534-273-7 paperback
978-1-59534-274-4 ebook

Note from the publisher: Great Salt Lake is ecologically important and faces complex challenges as it undergoes significant changes. Publication of this book was postponed due to complications from the COVID-19 pandemic, not delay on the part of the author. It is the publisher's hope that the book makes a timely and valuable contribution to the understanding, appreciation, and future of Great Salt Lake.

Trinity University Press strives to produce its books using methods and materials in an environmentally sensitive manner. We favor working with manufacturers that practice sustainable management of all natural resources, produce paper using recycled stock, and manage forests with the best possible practices for people, biodiversity, and sustainability. The press is a member of the Green Press Initiative, a nonprofit program dedicated to supporting publishers in their efforts to reduce their impacts on endangered forests, climate change, and forest-dependent communities.

The paper used in this publication meets the minimum requirements of the American National Standard for Information Sciences—Permanence of Paper for Printed Library Materials, ANSI 39.48-1992.

CIP data on file at the Library of Congress
27 26 25 24 23 | 5 4 3 2 1

CPSIA information can be obtained
at www.ICGtesting.com
Printed in the USA
JSHW041243200423
40551JS00001B/1

9 781595 342737